"What happens in practice is that by intuitive insight, or other inexplicable inspiration, the theorist decides that certain features seem to him more important than others and capable of explanation by certain hypotheses."

Fred Hoyle (1915-2001)

NEW DIMENSIONS IN ELEMENTARY PARTICLE
PHYSICS AND COSMOLOGY

NEW DIMENSIONS IN ELEMENTARY PARTICLE
PHYSICS AND COSMOLOGY

RESULTS of Independent Theoretical *RESEARCH*
INCLUDING A *SIMPLE MODEL* FOR THE
'GOD PARTICLE'

Second Edition

Ashok K. Sinha

To order additional copies of this book, contact:
Xlibris LLC
1-888-795-4274
www.Xlibris.com
Orders@Xlibris.com
124429

CONTENTS

LIST OF TABLES

LIST OF FIGURES

BOOKS BY ASHOK SINHA

IN ENGLISH

NEW DIMENSIONS IN PARTICLE PHYSICS AND COSMOLOGY

THEORY OF SATELLITE AND CELLULAR (MOBILE) TELECOMMUNICATIONS

DROPS OF DEW (Collection of Poems)

MANIFESTATIONS OF THOUGHTS (Verse Translation of a Sufi Poetical Work)

GAZALIATS (Verse Translation of a Sufi Poetical Gazals)

THE BATTLEFIELD OF KURUKSHETRA (Verse Translation of the Hindi Classic 'Kurukshetra'—of the Mahabharat War of India—by the Great Poet 'Dinakar')

ALEXANDER AND CHANDRAGUPTA THE GREAT (Historical Play)

SHAKUNTALA (Translation of the Sanskrit Classic Play by the Great Poet KALIDAS)

THE SUBLIME JOY OF THE GEETANJALIC PSALMODY (Translation of poems from the Collection "Geetanjali" by Nobel Laureate, Rabindra Nath Tagore)

THE NEXT LIFE (Novel with glimpses of India's past and present)

A BIRD'S EYE-VIEW OF THE SANATAN DHARMA (Hinduism) (with Dr. Shardanand)

RELIGIONS OF INDIAN ORIGIN: HINDUIS, BUDDHISM, JAINISM, SIKHISM (with Dr. Shardananand)

EVOLUTION AND DEVOLUTION OF THE CONCEPT OF GOD IN WORLD-RELIGIONS*(under preparation)

MODERN VIEWS AND MUSE ON HINDUISM (Editorials and Short Essays)

YOGA AND AFFILIATED PRACTICES AND SCIENCE* (under preparation)

IN HINDI

THE BHAGVAD-GEETA (Verse Translation of the Sanskrit Scripture)

PUNYADHANWA (Poetical Work Based on the Life of EKALAVYA)

BULBULON KE DARPAN ME (Collection of Poems)

INDRADHANUSH (Collection of Short Stories and One-Act Plays)

OMAR KHAYYAM KI RUBAIYAN (Translation of the 75 Well-Known Rubaya'ts of Omar Khayyam)

OMAR KHAYYAM KI NAI RUBAIYAN (Trans. of the 512 Lesser-Known Rubaya'ts of Omar Khayyam)

INTRODUCTION

This book introduces a new paradigm involving the concept of a three-dimensional time (3-D T) which, together with the usual three-dimensional space (3-D S), forms a six-dimensional spacetime (6-D ST) continuum for describing super-high energy (Planck-scale) elementary particle and cosmological phenomena. For a spherically symmetric system, this leads to the notion of a 5-sphere topology. In this framework, a heuristic model and a simple theory of elementary particles and the four basic interactions (strong, weak, electromagnetic and gravitational) is developed, including a simple representation of the elementary particle masses in terms of the basic parameters of these interactions. Emphasis is laid on mathematical simplicity and a direct use of elementary physics. For this reason, no reference to the string theory is made.

It should be noted that while the concept of 3-D T (6-D ST) and the quaternion algebra make the theory logically complete and impart useful physical significance to the the model, these elements are not essential to the model for the representation of the particle mass in terms of the interaction coefficients. This representation can thus be treated as a heuristic formula for extension of the model to include the Higgs bosons.

Content-Distribution

The book is divided into eight Chapters.

Chapter 1 introduces the concept of a three-dimensional time (i.e., a six-dimensional spacetime continuum.)

Chapter 2 applies this concept to develop a simple model of elementary particles in the framework of the basic interactions. This somewhat long Chapter summarizes most of the important results of analytical studies and mathematically simple modeling effort; with potential applications to cosmological 'problems,' albeit conceptually. The concepts of 'phase-changes' through rotations in the hyperspace, described in terms of quaternion algebra, are introduced. It is demonstrated that the particle mass can be interpreted as a directly correlated to the basic interactions.

Chapter 3 presents a simple matrix representation of the particle mass-interaction Relationship.

Chapter 4 develops a statistical theory of a relativistic ensemble of particles in the conventional 4-covariant and the new 6-covariant forms.

Chapter 5 and Chapter 6 present a unified representation of the four basic interactions involved, using algebraic and geometrical formulations, respectively, introducing a "spring theory" for the strong interaction in the former case.

Chapter 7 relates to certain selected aspects of basic physics, including a unified representation of Newton's familiar Laws of Motion, and interrelation between classical and relativistic mechanics.

Chapter 8 briefly discusses the origin of spin and related factors.

There are two Appendices at the end of the book.

Appendix A describes a cosmic time-scale based on age-old Hindu scriptures, showing how it agrees fairly well with the results of modern cosmology, including the age of the solar system. This cosmic time-scale, however, propounds a cyclic universe, with periodic 'Creation' and 'Dissolution' of 'Universes' in different epochs, in contrast to the hypothesis of 'Parallel Universes (Multiverse).'

Appendix **B** briefly lists the names, periods, and achievements of a few ancient Indian scientists. It is indicated that these scientists had discovered a number of the theories in physics, cosmology, mathematics, chemistry, medicine, surgery, etc., many of which are now ascribed to other, more recent scientists.

Physical Significance of the Extra Dimensions of Time

A word is due by way of a simple and natural explanation of the genesis of the three-dimensional time—a generalization of Einstein's special relativity theory (SRT) that may be considered not only desirable but indispensable. The SRT treats the relative 'velocity' between two reference frames in terms of a single component, effectively reducing it to a scalar. The argument that one can choose the orientations of the two reference frames with one common axis (say the common x-axis) parallel to the direction of the relative velocity may not always be tenable or valid, due to high energy phenomena, in elementary particle physics and in cosmology, involving rotational or rapidly changing motions, or the need to deal with multiple bodies with arbitrary velocities, simultaneously.

Methodology

Of course, there should be hardly any mystery or resistance to the concept of the physical relevance and interpretation of the two additional components of time—these refer to the motion along the corresponding coordinates: the longitudinal (azimuthal) time component refers to the component of motion in the longitudinal (azimuthal) angular direction in a spherical polar coordinate system. The inherent spherical symmetry in the universe makes this type of coordinate system most suitable for the purpose of describing the very small (sub-nuclear or elementary particle) and very large (cosmological) phenomena.

In a spherical polar coordinate system, the conventional "positive arrow of time" of common experience—associated with the increasing entropy—may be taken as the 'radial' component of a three-dimensional time vector, while the other two ('longitudinal' and 'azimuthal') components are postulated here to play important roles only in high-energy (elementary particle and cosmological) phenomena. Rotations in a four-dimensional space is most conveniently described by using the quaternion algebra, discovered by Hamilton in 1843. A quaternion contains one real (space) variable and three imaginary (time) variables, together called the four 'generators' of the quaternion. Thus, as in case of many other new developments in physics, the mathematics to handle the pertinent theory already exists beforehand in the present case, as well.

The two additional components of time may be considered as the hidden or 'curled' variables, their roles or effects becoming manifest only at super-high-energy range. Also the six-dimensional spacetime, together with the five interaction 'constants' of elementary particles (the parameters for the strong, weak, electromagnetic or electric charge, gravitational or mass, and spin), could be regarded to constitute an 11-dimensional space to describe the elementary particles, as is done in the superstring theory. However, the present theory is mathematically much simpler and without any ambiguity.

Applications

The questions of the dark matter and of the dark energy are also resolved in a natural fashion under the present theory. In short, one can say that they exist but are invisible to our observations, simply because they are not emitting radiation (dark matter) or the relevant motion is taking place in the angular direction, transverse to our direction of observation (i.e., the radial direction.) Furthermore, the Hubble law should automatically be expected to predict radial acceleration of cosmological bodies.

The theory presented here also admits of further extensions and generalizations. One such extension presented here is introduction of the Higgs interaction parameter, whose value is zero for the known elementary particles of Family 1, 2, and 3, but one for the *Higgs boson*. In the present framework, then, the masses of the three Higgs bosons corresponding to Family 1, 2, and 3 are 'predicted' here, one value, corresponding to Family 3, coinciding with the value estimated by the *CERN/LHC experiments and announced recently (July 4, 2012.)*

... ...

CHAPTER 1

A Unified Theory of Elementary Particles and Interactions (UTOEPI): The Concept of 3-Dimensional Time (6-D Spacetime)

> "Most of the fundamental ideas of science are essentially simple, and may, as a rule, be expressed in a language comprehensible to everyone."
>
> Albert Einstein (1879-1955)

1.1. Introduction

In this Chapter, a new paradigm is introduced involving the concept of a three-dimensional time (3-D T) which, together with the usual three-dimensional space (3-D S), forms a six-dimensional completely orthogonal spacetime (6-D ST) continuum for describing super-high energy (Planck-scale) elementary particle and (large-scale) cosmological phenomena. The need for an introduction of such a 3-D T (6-D ST) is highlighted in reference to the basic criteria of Einstein's theory of relativity. An application of this concept in modeling the elementary particles, including the Higgs bosons ('the God Particle') and the basic interactions thereof, as well as in cosmology, in a mathematically simple and unified manner, is the subject of this book.

We start with a discussion of why Einstein's theory of relativity based on a single time dimensions needs to be modified by introducing a three-dimensional (3-D) time, corresponding to the 3-D space, for an arbitrary 3-D vector for the relative velocity between two frames of references; and how the 3-D time components relate to the conventional (1-D) time.

This Chapter presents arguments to highlight the need to extend the concept of the relativistic four-dimensional spacetime (4-D ST) to a new paradigm of the six-dimensional space-time (6-D ST) continuum by introducing a three-dimensional time (3-D T)-space orthogonal to the conventional 3-dimensional 'real' space (3-D S), with the three time coordinates also forming an 'imaginary' right-hand orthogonal Cartesian reference frame similar to the 3-D S. Apart from the conventional time (associated with the "positive arrow" of time of increasing entropy, of common physical phenomena, and of every-day experience), the additional two orthogonal time-dimensions may be regarded analogous to the familiar Kaluza-Klein[1] ("curled-up") variables playing a significant role only in ultra-high energy and Planck-scale spacetime range pertinent only for sub-nuclear particles created in the big bang and in high-energy particle accelerator, for example, and for cosmological phenomena.

1.2. Extended Relativity Theory

In order to distinguish the concept and principle of 6-D ST from the conventional 4-D ST of the Special Theory of Relativity (SRT), we will refer to the 6-D ST-based theory as the "Extended Relativity Theory" (ERT.) The elementary particle and basic interaction model associated with the 6-D ST will be referred to as the universal and Unified Theory of Elementary Particles and Interactions ("UTOEPI".) Before proceeding to the presentation of the ERT and UTOEPI, it is useful to outline the need and justification for introducing the 3-D T (i.e., the 6-D ST) concept.

1.3.   The Need to Consider Three-Dimensional Time (3-D T)

The Lorentz transformation rules are derivable from the special relativistic invariance of the (4-D) space-time interval

$$(dx)^2 + (dy)^2 + (dz)^2 + (icdt)^2 = 0$$

where $i = \sqrt{(-1)}$, and c is the velocity of light.

Commonly, the relative velocity, v, with which a Cartesian coordinate frame, R', is assumed to move with respect to the reference Cartesian coordinate frame, R, is taken along one common coordinate axis, say, the x-axis, in order to obtain the well-familiar Lorentz transformation equations between the coordinates (x', y', z', t') of R' in terms of the coordinates (x, y, z, t) of R:

$$x' = (x - vt)\beta; y' = y; z' = z; \tag{1.1.a}$$

$$t' = \left(t - \frac{vx}{c^2}\right)\beta \tag{1.1,b}$$

$$\beta = \left(1 - \frac{v^2}{c^2}\right)^{-\frac{1}{2}} \tag{1.1.c}$$

The above treatment of the special relativity theory (SRT), explicitly assuming the relative velocity vector (v) to be aligned with the x-axis, is tantamount to reducing (v) to a scalar, i.e.,

$$\underline{v} = v_x \hat{i}, \ v_y = v_z = 0, |\underline{v}| = v_x , \tag{1.1.d}$$

where the suffixes, x, y, z, denote the components along the three axes (x, y, z), and $\hat{i}$ is the unit vector along the x-axis of R. In a more general case, however, the velocity v may be retained as a three dimensional vector with non-zero y—and/or z-components

$$v = v_x \hat{i} + v_y \hat{j} + v_z \hat{k} \tag{1.1.e}$$

$\hat{i}, \hat{j}, \hat{k}$ along the x-, y- and z-axes, respectively, forming a triad of orthonormal vectors in the Cartesian coordinate frame R:

$$\left(\hat{i}\right)^2 = \left(\hat{j}\right)^2 = \left(\hat{k}\right)^2 = 1$$

$$\hat{i} \cdot \hat{j} = \hat{j} \cdot \hat{k} = \hat{k} \cdot \hat{i} = 0, \tag{1.1.f}$$

where the dot denotes commutative scalar product. In such a general case ($v_y \neq 0$ and/or $v_z \neq 0$), it would be essential to introduce independent time-coordinates corresponding to the spatial coordinates y and z, respectively, in addition to the 'conventional time' t of the 4-D ST of the SRT. In the interest of generality, we now denote the (3-D) spatial coordinates as (x_1, x_2, x_3) and the corresponding (orthogonal) 3-D time coordinates as (ict_1, ict_2, ict_3), thereby obtaining a six-dimensional space-time (6-D ST) continuum (x_k, ict_k, k = 1, 2, 3); or a 6-D ST hyperspace, alternatively also denoted as having the 6 orthogonal coordinates (x_1, x_2, x_3, x_4, x_5, x_6), with the notations: $x_4 = ict_1$; $x_5 = ict_2$; and $x_6 = ict_3$. For a relative velocity vector v with three orthogonal components (v_1, v_2, v_3) along the three orthogonal coordinate spatial axes (x_1, x_2, x_3), respectively, of R, if a scalar (one-dimensional) time variable, t, is employed, as implicitly implied in the construct of Einstein's four dimensional space-time (4-D ST) continuum, serious ambiguities, contradictions and inconsistencies result. This is readily seen by noting that the set of equations

$$x_i' = \left(x_i - v_i t\right)\beta_i \tag{1.2.a}$$

$$t' = \left(t - \frac{v_i}{c^2} x_i\right)\beta_i \tag{1.2.b}$$

$$\beta_i = \left(1 - \frac{v_i^2}{c^2}\right)^{-\frac{1}{2}}, \quad i = 1, 2, 3 \tag{1.2.c}$$

requires that the identity

$$t' \equiv \left(t - \frac{v_1}{c^2} x_1 \right) \beta_1 \equiv \left(t - \frac{v_2}{c^2} x_2 \right) \beta_2 \equiv \left(t - \frac{v_3}{c^2} x_3 \right) \beta_3 \qquad (1.2.d)$$

must hold: a condition that disallows an arbitrary choice for the velocity components (v_1, v_2, v_3). Similarly, it can be easily verified that the basic invariance relation of the 4-D ST

$$x_1'^2 + x_2'^2 + x_3'^2 - (ct')^2 = x_1^2 + x_2^2 + x_3^2 - (ct)^2 \qquad (1.2.e)$$

becomes untenable, violating the fundamental premise of SRT. The above observations amply explain the need for, and justify, the introduction of 3-D time (6-D timespace) for the sake of generality, to adequately describe the case where at least two of the velocity components v_x, v_y, and v_z are non-zero with respect to the observer. For such a general case, then, the general Lorentz transformation equations are provided by the following equations:

$$x_i' = \left(x_i - v_i t_i \right) \beta_i \qquad (1.3.a)$$

$$t_i' = \left(t_i - \frac{v_i}{c^2} x_i \right) \beta_i, \; i = 1, 2, 3 \qquad (1.3.b)$$

With the resulting 3-D T (hence, 6-D ST) concept postulated to hold in general, there is no contradiction or ambiguity and the invariance relation

$$\sum_1^6 \left[(x'_i)^2 - (x_i)^2 \right] = 0, \qquad (1.3.c)$$

basic to the theory of relativity, is satisfied. This establishes the validity of the extended relativity theory (ERT) characterized by a multidimensional velocity vector, having two or three non-vanishing components with respect to the observer.

Clearly, the physical implications of the above discussion are obvious and significant. Of course, in practice, these implications and theoretical modifications are relevant only for high energy domain, when the transverse velocity components with respect to the observer are large, approaching the velocity of light. For this reason, the ERT (6-D ST)-based analysis and treatment of observed results should prove to be more relevant and appropriate than SRT mainly in the context of cosmological and high-energy particle accelerator events. Creation of, and interaction among, elementary particles during the big bang, obviously constitute prime examples of such events. Of special interest is the potential application of the 6-D ST concept in building model(s) and a Unified Theory of Elementary Particles and Interactions (UTOEPI), the subject of this article. The four basic interactions are known to be inherently spherically symmetric. Let us, therefore, briefly summarize the mathematical method to describe spherical symmetry in six-dimensional orthogonal space-time continuum (6-D ST) before exploring the associated system properties

1.4. Spherical Symmetry in the 6-D ST Hyperspace

In a spherically symmetric system, the radial distance, r, of a point from the origin of the reference frame, or between two points, is the only variable that explicitly governs the basic interaction pattern and the resulting behavior of the particles and fields. In other words, spherical symmetry equivalently implies system invariance, and hence inherent symmetry, with respect to angular coordinates designating the orientation of the radius vector or of the straight line joining the interacting particles or bodies,

$$r = \left(x_1^2 + x_2^2 + x_3^2 \right)^{\frac{1}{2}}$$

and, analogously, in the time-domain, we can define an analogous 'radial time-separation'

$$ict_r = \left(x_4^2 + x_5^2 + x_6^2\right)^{\frac{1}{2}}$$

For certain applications, it may be useful, for the sake of simplicity, to consider a 2-dimensional hyperspace (i.e., a "hyper-plane") constituted of the two "non-negative" coordinates, $r \geq 0$ and $t_r \geq 0$, the latter denoting the conventional time coordinate, associated with the "positive arrow" of time.

The well-known fact that the strong interaction between two quarks increases with increasing separation between them naturally allows a simple modeling of the same in terms of a "spring"-like force or potential; and the related formulation of the strong interaction may be referred to as the '*spring theory*.' The electromagnetic (EM) and gravitational interaction potentials vary in inverse proportion to the radial separation. Thus the radial separation between two points, specifying the locations of the two interacting systems with respect to the observer, is of prime interest in most cases.

In case of a cosmological event, the variation in the location of the observer can obviously be considered relatively negligible due to the 'astronomical' distances involved; whereas it may play a critical role in the case of particle-accelerator related observations.

CHAPTER 2

Potential Applications of the 3-D Time (6-D Spacetime) Concept

". . . Space is not a lot of points close together; it is lot of distances interlocked"

"Let us draw an arrow arbitrarily. If as we follow the arrow we find more and more of the random elements in the state of the world, then the arrow is pointing towards the future; if the random element decreases the arrow points towards the past... I shall use the phrase 'time's arrow' to express the one-way property of time which has no analogue in space."

Arthur Stanley Eddington (1882-1944)

"$i^2 = j^2 = k^2 = ijk = -1$

These formulae were conceived on 16th October 1843, and carved by Hamilton, apparently in this form on a stone of Brougham Bridge, over the Royal Canal, Dublin, at the time."

William Rowan Hamilton (1805-1865)

Before considering the modeling of the system of elementary particles, it is interesting to illustrate potential applications of the 6-D ST concept. Examples of the prominent unresolved questions of cosmology and particle physics include, for example, (a) What is the exact nature of the "positive arrow" of Time? (b) What constitutes the dark matter? (c) What is the source of the dark energy? These 'problems' are considered in Chapter 2 qualitatively in the light of the 6-D ST concept to illustrate how this concept could help resolve them.

2.1. The Dark Matter

We shall consider the validity and extension or generalization of the big bang concept, in terms of cyclic universe and / or multi-verse concepts later in this book. Here we start with the conventional big bang concept for simplicity.

It is reasonable to assume that the primordial matter manifested in the big bang, or existing at the present epoch of cosmological evolution, is likely to be grossly distributed in equal proportions in all dimensions of space-time continuum, since, *ab initio*, no particular direction or dimension inherently enjoys especial distinction or preference by nature. We refer to this hypothesis as the *'Dimensional Equi-Partition of Matter (DEM).'* The time-dimensions could also be equivalently represented by the momentum space, since their direct relevance is associated with motion or momentum of particles or bodies—without any change due to momentum-changes, time loses much of its significance. Thus the 6-D ST is, for many purposes, equivalent to the phase-space of thermodynamics, composed of three position-coordinates and three momentum coordinates.

Under the assumption of DEM, the amount of matter attributable to each single dimension of the phase-space, expressed as a percentage of the total amount of matter existing in the universe as a whole, can be simply written as

$$P_1 = 100 / N \%, \tag{2.1a}$$

where N is the total number of independent (orthogonal) dimensions. Setting N = 6, we obtain

$$P_1 = 100 / 6 \% \approx 16.67 \% \tag{2.1b}$$

Now, we recall that only Family 1 (F_1) elementary particles undergo formation of ordinary matter consisting of atoms and molecules, which possess characteristic set of energy levels and thereby provide the possibility of specific patterns of emission of radiation in various spectral ranges due to transitions from a higher energy level to a lower one. The elementary particles of Family 2 (F_2) and Family 3 (F_3) essentially behave like inert matter, and simply do not evolve to form atoms and molecules. Therefore the F_2 and F_3 particles can make no contribution in spectral radiation in the universe. It is also reasonable to assume that equal amount of matter can be attributed to each of the three Families of elementary particles. We refer to this hypothesis as the '*Family Equi-Partition of Matter (FEM).*' Under this assumption, the percentage of matter that emits radiation, traveling in one particular direction (i.e., attributable to a single spatial dimension in the 6-D ST), is then only one-third of the value of P_1 given above. It is radiation alone that is responsible for rendering such matter optically visible through earth- and space-based optical telescopes (operating in various parts of the spectrum including ultra-violet, visible and infra-red ranges.) Therefore, the percentage, P_v, of such 'visible matter,' in comparison to the net amount of matter present in the universe, can be roughly estimated to be given as

$$P_v = P_1 / 3 \approx \left(16.\dot{6} / 3\right) \% \approx 5.56\% \tag{2.1c}$$

Clearly, P_v provides an upper limit of the theoretically estimated percentage of the visible matter in the universe, and is consistent with the estimated value from observational data. The above result also indicates that the mysterious dark matter in the universe can be identified with the large amount of inert matter of Family 2 and Family 3 elementary particles around galaxies and galaxy clusters. Such inert products of the big bang, practically unchanged throughout the age of the universe and pervading the universal 6-D space-time continuum, might in fact have played a significant role in the formation and evolution of galaxies and galaxy-clusters composed of visible matter, by virtue of their large gravitational interaction effects.

2.2. The Dark Energy

Einstein's formula for energy, $E = mc^2$, is obtained by integrating the kinetic energy of a mass over all possible velocity values ($0 \leq v \leq c$). However, in this formula for energy, velocity is treated as one-dimensional (scalar) quantity, as mentioned above. If all three possible velocity components are considered, as assumed in this book, we would have other two orthogonal components of velocity also yielding corresponding energy values, each equal to mc^2. For cosmological bodies, these other two components comprise the dark matter, as explained above. Hence, these 'other' two portions of energy ($2\ mc^2$) not being observable due to their correspondence with the angular components of the time-vector, normal to the direction of observation, can be said to be, or contribute to, the so-called dark energy. Only one portion of energy, corresponding to the velocity component aligned with the direction of observation, is detected, which is only one-third of the total energy ($3\ mc^2$) that can be attributed to the body in question. The remaining energy, twice in magnitude ($2\ mc^2$) remains undetectable. Additional contribution to the dark energy of course is associated with the dark matter itself, and can be estimated as $3M_dc^2$, where M_d is the amount of the dark matter present in the universe, or in any segment of it (galaxy, cluster of galaxies.)

It may be noted in passing that radial acceleration may also be naturally expected under the familiar Hubble Law. Thus, denoting the radial velocity of a galaxy at a distance R from the earth as V_R, we have

$$\Delta R/\Delta t_r = V_R = H\ R \tag{2.2a}$$

where H is the Hubble constant. Differentiating with respect to the radial time, t_r, we get

$$\Delta^2 R / \Delta t_r^2 = \Delta V_R / \Delta t_r = H \, \Delta R / \Delta tr$$

$$= H \, V_R$$

$$= H \, (H \, R) = H^2 \, R \qquad (2.2b)$$

In other words, the galaxy has a radial velocity as well as a radial acceleration which is essentially required by the Hubble Law, so that the galaxy assumes the new, higher velocity as it attains a new radial distance from the earth in course of its rapid outward motion away from every other galaxy under the Hubble expansion of the universe. Also, obviously, the magnitude of the acceleration itself is increasing (that is, there is progressively higher order of acceleration involved) as the galaxy continually moves in course of this expansion, as may be verified by successive partial differentiation of both sides of the above equation with respect to (radial) time. Of course, there may be additional motion including similarly various order of acceleration in the corresponding spatial directions, obtainable by partial differentiation with respect to the longitudinal and azimuthal time coordinates, respectively, which earth-based observations are unable to detect or to measure.

2.3. The Expected Number of Elementary Particles and of the Basic Interactions

We review the so-called Standard Model of elementary Particles later in this Chapter. Here we first examine the expected number of elementary particles and the expected number of their basic interactions.

2.3.1. Number of Elementary Particles

At this juncture, it is useful to enumerate the total number of (the Standard Model) elementary particles to be expected on the basis of the above characterization and 6-D ST scheme for their description. The self-evident spherical symmetry characteristics, leading to the 5-sphere (see Sec. 2.7) based description of the elementary particles, simplifies this process, making the determination of the total number of "all possible" elementary particles rather trivial. In how many ways could a set of five independent parameters be divided in three separate families? Clearly, the answer is the combination $^5C_3$.

$$^5C_3 = \frac{5!}{3!(5-3)!} = 10 = N(say)$$

$$(2.3a)$$

Based on the Standard Model of elementary particles, there are three Families (F_1, F_2, and F_3) of these particles, with each Family containing four particles, making the total number of elementary particles as 12 (see Table 2.1). However, experimental evidences indicate that the tree neutrinos, v_e, v_μ and v_τ, respectively, exhibit a special property, referred to as the "mixing" or "entanglement" phenomenon, according to which these three 'elementary particles' could be considered as the morphological manifestations of super-positions of three 'color' states of two and the same particle. In other words, any one of the three neutrinos, v_e, v_μ, and v_τ, could be regarded as altered forms or states of the third neutrino itself. Thus, as far as counting the number of independent elementary particles is concerned, under the above view-point, the total number of elementary particles should be taken as

$$N = 12 - 2 = 10 \qquad (2.3b)$$

This conforms to the expected number of elementary particles based on the hypothesis of 6-D ST hyperspace (i.e., a 5-sphere representation for the spherically symmetric case applicable to the elementary particles) as shown above. Again, the Family 1 (F_1) elementary particles, stable and directly relevant for the structure of the observable 'normal' universe, could be associated with the 'positive' sense (spherical-radial) time coordinate t_r, while the other two Family (F_2, F_3) elementary particles could be assumed to be associated with the (spherical-angular) time coordinates t_θ and t_ϕ. This aspect of the (spherical) time coordinates is further discussed below.

2.3.2. Number of Basic Interactions

Finally, the expected number of the interaction carriers could be determined by recognizing that the four members of Family 1 of the elementary particles interact on a two-by-two basis. Thus the total number of such carriers, within the framework of the Standard Model, is expected to be equal to the binomial coefficient $^4C_2 = 6$. We indeed have six bosons as interaction carriers, viz:

the gluon (for the strong interaction); the three weak bosons: $W^\pm$ and Z (for the weak interaction); the photon (for the EM interaction); and the graviton (for the gravitational interaction.) The interaction pattern for Family 1 $(F_1 (= F1))$ particles is illustrated in Figure 2.1.

It is interesting to note that, in view of the unique role of the Family 1 particles as being solely responsible for all the commonly *visible* matter in the universe, it may be said that the ten objects (4 elementary particles and 6 interaction carriers pertaining to Family 1) indicated in Figure 2.1 account for the entire *directly observable* universe with its myriad of manifestations, including all biological species and '*visible*' cosmological bodies. A simple model of the three Families of elementary particles and their interactions, leading to an expression for the particle mass in terms of their constant interaction parameters, will be presented later.

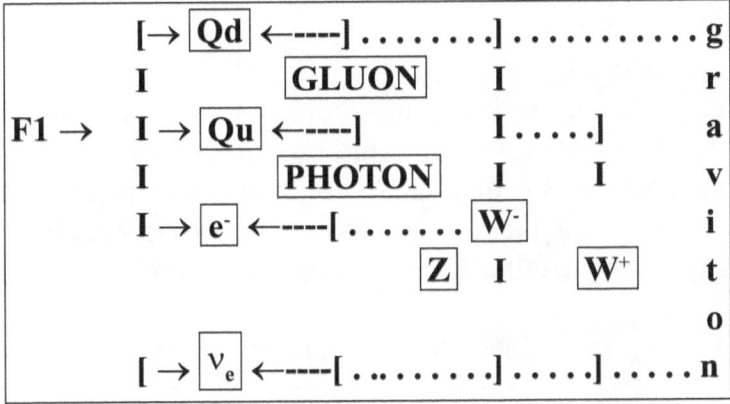

Figure 2.1.   The six Interaction Carriers (the gluon, the photon, the 3 weak bosons: $W^\pm$ and Z, and the graviton) illustrating the interaction pattern within the Family 1 (F1) elementary particles.

2.4.   The Positive Arrow of Time

Just as the spatial radial distance (r) represents a non-negative spatial distance from the origin of a points in space with spatial orthogonal Cartesian coordinates (x_1, x_2, x_3), the analogous radial "time distance" (t_r) represents a non-negative component of the associated event from the origin in the orthogonal time-Cartesian coordinates i.e. (omitting the factor 'ic' for brevity), (t_r, t_θ, t_ϕ) = (x_4, x_5, x_6) in the 3-D space-time (and hence in the 6-D ST hyperspace); the variable $x_4 = t_r$, therefore, is identifiable with the so-called "positive arrow" of time, associated with increasing entropy of a closed system in consistence with the Second Law of Thermodynamics ("Entropy always increases with time.") In the 6-D ST hyperspace hypothesized in this book, however, we postulate that there are additional (orthogonal) time coordinates (t_θ, t_ϕ) or (x_5, x_6). Considerations of the geometry and topology of the relativistic light-cone in the 6-D ST, in the context of the past, present and future time in the ERT, should be an interesting exercise.

It is useful to point out here that an alternative mathematical treatment of the combination of the radial space coordinate (r) and the three time coordinates, in the form of a quaternion can also be considered.

Specifically the radial time coordinate, $t_r = t_1$ (say), always having a non-negative value, could be interpreted as representing the positive arrow of time ($t_r \geq 0$), while the other two components ($t_\theta = t_2$ and $t_\phi = t_3$) could assume positive or negative values depending on the orientation of the time-vector t.

2.5. The Standard Model of Elementary Particles

2.5.1. Summary of Particle and Field (Interaction) Parameters

Table 2.1 summarizes the values of the five parameters for each well-known elementary article of the Standard Model. These parameters are arranged in three conventional groups or "Families" of elementary particles, F_1, F_2 and F_3, respectively, on the basis of experimentally observed characteristics of these elementary particles, including values of mass (M), spin (s), strong-interaction parameter (S), weak-interaction parameter (w), and electric charge (q). The degenerate 3 M-values (for identical s, S, w, q values in F_1, F_2, F_3, respectively, characterize the gravitational interaction, while the q-value, the electromagnetic (EM) interaction. Here we use the symbols

Q_u (up quark); Q_d (down quark); Q_c (charm quark); Q_s (strange quark); Q_t (top quark); Q_b (bottom quark); e^- (electron); μ^{-1}(muan); τ^{-1}(tau); v_e (electron neutrino); v_μ (muan neutrino); v_τ(tau neutrino).

TABLE 2.1—THE STANDARD MODEL:
Parameter Values of Elementary Particles
(Particle mass is in units of proton-mass,
$M_p = 1851.85 M_e$; M_e = electron - mass = 0.5Mev)

	Particle	Mass (M)	Strong (S)	Weak (w)	EM (q)	Spin (s)
Family 1: (F_1)	Q_u	$M_u = 0.0047$	1	½	⅔	½
	Q_d	$M_d = 0.0074$	1	½	-⅓	½
	e^-	$M_e = 0.00054$	0	-½	-1	½
	ν_e	$M_{\nu e} \leq 10^{-8}$	0	-½	0	½
Family 2: (F_2)	Q_c	$M_c = 1.6$	1	½	⅔	½
	Q_s	$M_s = 0.16$	1	½	-⅓	½
	μ^1	$M_\mu = 0.11$	0	-½	-1	½
	ν_μ	$M_{\nu\mu} \leq 0.0003$	0	-½	0	½
Family 3: (F_3)	Q_t	$M_t = 189$	1	½	⅔	½
	Q_b	$M_b = 5.2$	1	½	-⅓	½
	τ^1	$M_\tau = 1.9$	0	-½	-1	½
	ν_τ	$M_{\nu\tau} = 0.33$	0	-½	0	½

2.6. The Peculiarity of Family 2 and Family 3 Particles: The 'Dark' Secret

The interpretation of t_r as the commonly observable (conventional) time and its possible association with the Family 1 elementary particles is based on the fact that the Family 1 particles adequately account for the directly observable, 'real-world' matter in the universe. In contrast; the Family 2 and Family 3 particles appear to be rather redundant, since they do not occur naturally as directly observable or stable matter. They appear only in fleeting glimpses in high-energy experimental observations in particle accelerators, decaying rapidly into the relatively more stable elementary particles of Family 1 (F_1). This suggests an association of the Family 2 and Family 3 particles with the two 'angular' time dimensions. Just as the angular time components are not part of ordinary observations and day-to-day experience, so also the associated elementary particles (Family 2 and Family 3) are presumably remote from ordinary observations.

2.7.   Application of the 6-D Space-time Concept to Elementary Particle Modeling

In this Section, a simple model for representing the masses of the Standard Model elementary particles is presented in terms of functions of the field or interactions (strong, weak, and electromagnetic) parameters. Using the quaternion formulism and 6-dimensional hyperspace, a physical, self-consistent model for the particle mass is generated which could be further generalized toward incorporating the concepts of the Higgs field and supersymmetry, for instance.

In a three-dimensional space (3-D S), only two variables—the longitudinal and azimuth angular coordinates θ and ϕ, respectively—are required to uniquely specify any point on the spherical surface of a sphere of given radius r. Similarly, by extension, in an orthogonal hyperspace of dimension n, only (n-1) independent (orthogonal) "coordinates" (independent parameters) are required to specify any point uniquely on the surface of the hyper-sphere of a given "radius" (which is, of course, the nth coordinate or parameter.) The hyper-sphere in question, in the n-dimensional hyperspace, is simply referred to as the (n-1) sphere. In particular, in the six-dimensional space-time (6-D ST) hyperspace referred to above, any point on the surface of a 5-sphere, of a given "radius," is uniquely specified with the help of 5 orthogonal (independent) parameters.

In the case of elementary particles, all basic interactions depend only on the "radial distance" (in the 6-D ST hyperspace), and hence the spherically symmetric (polar) coordinate system could be beneficially employed for the description of elementary particles. Thus, we are led to the consideration of a 5-sphere, i.e., a 5-dimensional hyper-surface, of which any point in uniquely specified by using five independent parameters. For the present purpose of developing a simple model of the elementary particles' masses in terms of the parameters of the basic interactions—strong, weak, electromagnetic (EM) and gravitation, in the framework of the Standard Model, in the 6-D ST hyperspace, we identify these five parameters, as: (1) *mass* (M); (2) *strong* charge (S); (3) *weak* charge (w); (4) *electric* (EM) charge (q); and (5) *spin* (s).

The spin may be regarded as in intrinsic property of the particle. All the elementary particles in the Standard Model have a common value of spin (s = ½), as these are all fermions. As mentioned above, the remaining parameters—strong charge (S), the weak charge (w), the electric (EM) charge (q) and the mass (M)—are obviously associated with the strong, weak, EM, and gravitational interactions (fields), respectively. For simplicity, at this stage, the origin and effect of a fifth plausible field—the Higgs field—and potentially related factors will be briefly discussed later.

As regards the coordinate variables, the basic interactions in particle physics and cosmology are known to essentially depend only on the "radial distance" in the 3-D S, and hence also in the 6-D ST hyperspace. Thus, by definition, these systems are spherically symmetric. The spherical symmetry lends especial advantage in the use of the spherical polar coordinate system. Thus, we are led to the consideration of a 5-sphere, i.e., a 5-dimensional hyper-surface, of which any point in uniquely specified by using five independent coordinates with respect to which the system is symmetrical, i.e., invariant. In other words, there exists a five-fold symmetry in this (6-D ST) hyperspace. In accordance with the well-known Noether's Theorem[2] stating that every degree of symmetry corresponds to a conserved parameter, there must exist five conserved parameters for each elementary particle, one for each dimension of symmetry. The five constant parameters are readily identifiable with the familiar interaction parameters uniquely associated with these elementary particles in the framework of the Standard Model, viz., (1) the *spin* (s); (2) *the strong* charge (S); (3) the *weak* charge (w); (4) the *electric (EM)* charge (q); and (5) the *gravitational charge*, identified here with the *mass* (M). This observation justifies the introduction of the 6-D ST, or 3-D T space concept, rendering it valid as a physical reality.

2.8.   Quaternion Formulation for Parametric Modeling of Elementary Particles

To develop a simple mathematical modeling for the proposed Unified Theory Of Elementary Particles and Interactions (UTOEPI), we note that the *'real'* (spatial) coordinate, r, and the three *'imaginary'* (time) coordinates, x_4, x_5 and x_6, of the 6-D ST naturally lend themselves to a treatment of the system in terms of a quaternion[2].

While the special relativity theory (SRT) is exclusively based on a single time coordinate ($x_4 = ict = ict_r$; $i = \sqrt{-1}$), in the extended relativity theory (ERT) presented here, we have two additional time coordinates—$(t_\theta, t_\phi) \Leftrightarrow (x_5, x_6) = (jct_2, kct_3)$, with corresponding conserved parameter values in the spherically symmetric system (the 5-sphere)—where we have introduced the additional symbols j and k (not to be confused with the unit vectors) analogous to i ; viz. $j = k = \sqrt{-1}$.

Together with the real variable r, the three orthogonal, imaginary time coordinates in the 6-D ST can be said to be associated with the *generators (i, j, k)* of a *quaternion* with the following *defining* algebraic relationships:

$$i^2 = j^2 = k^2 = -1 = i\,j\,k \qquad\qquad (2.4.a);$$

$$ij = k = -ji \;\; ; \;\; jk = i = -kj \;\; ; \;\; ki = j = -ik \qquad (2.4.b,c,d)$$

A quaternion Z and its magnitude can in general be represented in the form:

$$Z = a_0 + a_1 i + a_2 j + a_3 k; \tag{2.4.e}$$

$$|Z| = \left(a_0^2 + a_1^2 + a_2^2 + a_3^2\right)^{1/2} \tag{2.4.f}$$

where a_0, a_1, a_2 and a_3 are real.

Just as the rotation of a vector in the 3-D space (3-D S) can be represented in terms of the three Pauli spin matrices, the rotation of a vector in the 4-D space can be represented in terms of the quaternion generators. Also, for certain applications, it may be useful, for the sake of simplicity, to consider a 2-dimensional hyperspace (i.e., a "hyper-plane") constituted of the two "non-negative" coordinates $r \geq 0$ and $t_r \geq 0$, the latter denoting the conventional (Einsteinian) time coordinate, associated with the positive arrow of time, as already mentioned. It is readily evident that the Lorentz transformation or the Einstein's equations for the space-time coordinate transformations are confined to this (r, t_r) hyper-plane with the time coordinate orthogonal to the radial space coordinate which, by default, essentially constitutes the direction of observation. The present representation using the complete 6-D ST hyperspace thus reduces to the 4-D (Einsteinian) spacetime

$$(x_1, x_2, x_3, x_4) = (x, y, z, ict)$$

if the physical events in a spherically symmetric system are described in terms of the 2-D hyper-plane (r, t_r).

2.9.   Qualitative Discussion of the Quaternion Representation

To derive the desired relations among the interaction parameters taken as the basic properties of each elementary particle, recall that successive rotations in the 6-D ST, i.e., in the 4-D hyperspace $(r, ict_r, jct_\theta, kct_\phi)$, are mathematically represented by the generators (i, j, k) of a quaternion obeying the non-commutative algebra specified in Eqns. 2.4(a-d). The non-commutative properties of the quaternion reflect the analogous characteristics of the rotation operation: A rotation X followed by another rotation Y is *not* equal to the rotation Y followed by the rotation X ($XY \neq YX$).

The final grouping of the elementary particles of the Standard Model resulting from the 'spectral splitting' in the 3-D orthogonal space of the conserved interaction parameters (S, w, q) is schematically depicted in Figure 2.2. This splitting process in fact leads to four distinct 'Groups' denoted here as G_1, G_2, G_3, and G_4, respectively. Each Group comprises 3 particles having identical parameter values. This feature is exploited below toward further quantitative analysis, and can be said to form the basis of the simple model (UTOEPI) presented here. The three 'degenerate' mass-values within each Group correspond to the elementary particles of the three Families, F_1, F_2, and F_3, respectively, of the Standard Model. The 'Group' of each particle is shown in the last column of Table 2.2(A), which is simply a slightly modified form of Table 2.1.

TABLE 2.2(A)—THE STANDARD MODEL:
Parameter Values of Elementary Particles
(Particle mass is in units of proton-mass,
$M_p = 1851.85M_e$; M_e = electron - mass = $0.5M_{ev}$)

	Particle	Mass (M)	Strong (S)	Weak (w)	EM (q)	Group
Family 1: (F₁)	Q_u	$M_u = 0.0047$	1	½		G_1
	Q_d	$M_d = 0.0074$	1	½	-	G_2
	e^-	$M_e = 0.00054$	0	-½	-1	G_3
	ν_e	$M_{ve} \leq 10\text{-}8$	0	-½	0	G_4
Family 2: (F₂)	Q_c	$M_c = 1.6$	1	½		G_1
	Q_s	$M_s = 0.16$	1	½	-	G_2
	μ^{-1}	$M_\mu = 0.11$	0	-½	-1	G_3
	ν_μ	$M_{v\mu} \leq 0.0003$	0	-½	0	G_4
Family 3: (F₃)	Q_t	$M_t = 189$	1	½		G_1
	Q_b	$M_b = 5.2$	1	½	-	G_2
	τ^{-1}	$M_\tau = 1.9$	0	-½	-1	G_3
	ν_τ	$M_{v\tau} = 0.33$	0	-½	0	G_4

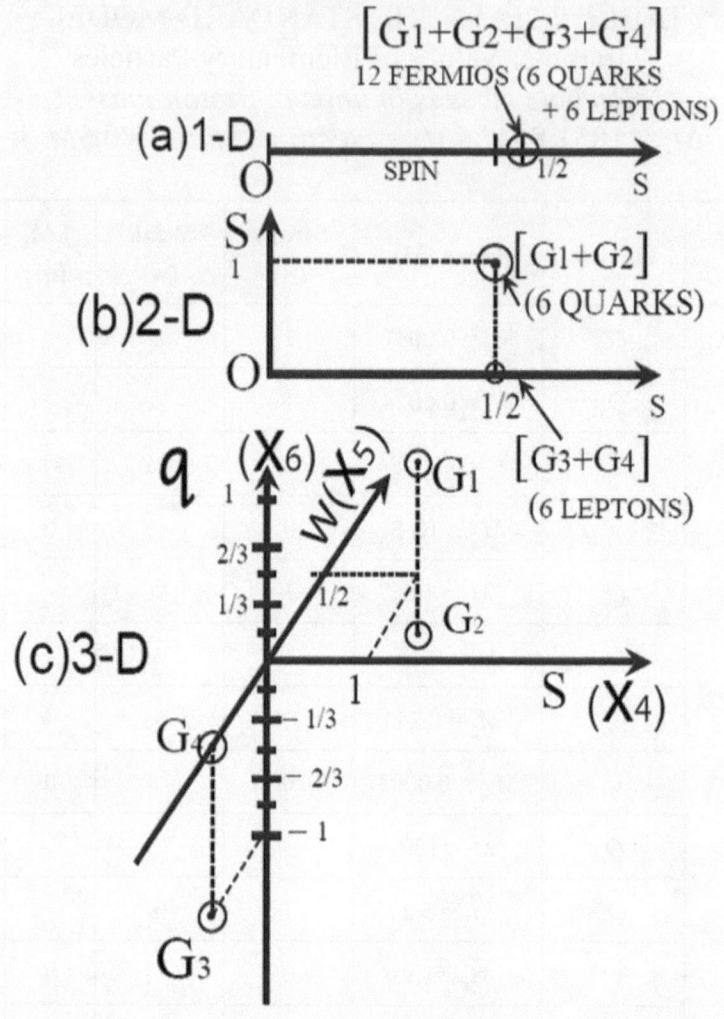

Figure 2.2.   Graphical representation of the four Groups (G_1, G_2, G_3, G_4) of the Standard Model elementary particles (a) in the 1-D space of spin (s=1/2) space; (b) in the 2-D space of the spin and strong interaction parameter (s=1/2, S=1); (c) in the 3-D space of the (S, w, q)-parameters represented along the (x_4, x_5, x_6) axes, respectively:

2.10. Quantitative Analysis and Implications

Further interrelations between the mathematical description and physical behavior of the system will be indicated below, providing a basis for the physical mass of the elementary particle to be expressed in terms of the associated interaction parameters. For this purpose, let us introduce a general quaternion, M_α, with the generators $(i, j, k,)$ as follows:

$$M_\alpha = \alpha_0 s + i\alpha_1 S + j\alpha_2 w + k\alpha_3 q; \quad (\alpha_0, \alpha_1, \alpha_2, \alpha_3 \text{ real}) \quad (2.5a)$$

By definition, the magnitude of this quaternion is given by

$$|M_\alpha| = \left[(\alpha_0 s)^2 + (\alpha_1 S)^2 + (\alpha_2 w)^2 + (\alpha_3 q)^2 \right]^{\frac{1}{2}} \quad (2.5b)$$

Since the real coefficients $(\alpha_0, \alpha_1, \alpha_2, \alpha_3)$ are arbitrary, as a first approximation let us simply assume: $\alpha_0 = \alpha_1 = \alpha_2 = \alpha_3 = 1$ and denote the corresponding value of the magnitude of the quaternion as M_Q: i.e.,

$$M_Q = \left[s^2 + S^2 + w^2 + q^2 \right]^{\frac{1}{2}} \quad (2.6a)$$

Clearly, this assumption could be interpreted as an assumption of equal strengths of the strong, weak and EM interactions. This represents the phase of the so-called Grand Unification following the big bang. This phase of the cosmological evolution clearly corresponds to extremely high energy- and temperature-levels, prior to the splitting of the three interactions under cosmological expansion and radiation-cooling. In this phase, the curves representing the variations of the interaction potentials as functions of the radial separations between the interacting particles in the 6-D ST converge into one single point (representing the Planck scale.) Incremental increase in the values of $\alpha_0, \alpha_1, \alpha_2, \alpha_3$ in quantum (integral or discrete) units may be examined to develop a simple theory of quantum gravity; whereas use of the interaction potentials as continuous functions of the radial separation could provide a classical theory of a unified representation of the four basic interactions involved.

The values of the basic parameters of the elementary particles are summarized in Table 2.2(B) where—instead of grouping the particles by the conventional Family-structure comprising Family 1 (F_1), Family 2 (F_2) and Family 3 (F_3), each Family containing four particles—the grouping is based on the above-mentioned Group-structure comprising the four groups, G_1, G_2, G_3, and G_4. Each Group thus contains three elementary particles with identical values of s, S, w, and q, and, hence, occupies the same position in the (S,w,q)-diagram (Figure 2.2). This of course reflects the inherent 5-fold symmetry in the corresponding 6-D ST coordinate (5-sphere) hyperspace. The three particles in each Group, in turn, reflect an inherent symmetry or degeneracy. Recall that this symmetry is broken by the introduction of the gravitational interaction, resulting in 'splitting' of each of these states into three degenerate mass-states; i.e., three separate elementary particles, belonging to the three Families F_1, F_2, and F_3, respectively.

For convenience in graphical representation of the elementary particle mass M (with numerical values spanning many decades) as a function of the interaction parameters, let us omit the spin-value, which is a common and constant ($s = \frac{1}{2}$) element for all the particles considered. This corresponds to the following arbitrary choice of the coefficients: $\alpha_0 = 0$, $\alpha_1 = \alpha_2 = \alpha_3 = 1$, still representing a phase of equal strengths for the three basic interactions as mentioned above.

The square of the magnitude of the corresponding quaternion, denoted as M_0, is then the sum of the squares of the strong charge (S), weak charge (w) and electric (EM) charge (q), and could be viewed as the square of the 'distance' from the origin to the point of interest in the orthogonal '(S, w, q)-space' (Pythagoras theorem):

$$M_0 = S^2 + w^2 + q^2 \tag{2.6b}$$

The values of M_0, normalized to a common denominator (36), are shown in Table 2.2(B); and the values of the logarithm of the particle mass in MeV, M_m:

$$M_m = \log_{10}(M) \tag{2.6c}$$

plotted against M_0 are shown in Figure 2.3. This plot, again consisting of three separate curves corresponding to the particles of the three Families (F_1, F_2, and F_3), shows an approximately linear dependence of the mass upon M_0, except for the case of Family 1, for which the mass of the up-quark conspicuously deviates from such a linearity. This deviation could, in fact, be regarded as fundamentally significant for the evolution of the universe, as discussed later in more detail.

TABLE 2.2—Elementary Particle Mass* and Interaction
Constants (The Standard Model) with
Group-wise Classification (G_1, G_2, G_3, G_4)

Group	Elementary Particle	Mass M (in MeV)	Basic Interaction Constants				$M_0 = (S^2 + w^2 + q^2)$
			s	S	w	q	
G_1	Qu	1.5-3.3	1/2	1	1/2	2/3	61/36
	Qs	$(1.16-1.34) \times 10^3$					
	Qt	$(1.69-1.73) \times 10^5$					
G_2	Qd	3.5-6.0	1/2	1	1/2	-1/3	49/36
	Qc	$(0.70-1.30) \times 10^2$					
	Qb	$(4.13-4.37) \times 10^3$					
G_3	e-	0.511	1/2	0	-1/2	-1	45/36
	u-	1.0571×10^2					
	t-	1.777×10^3					
G_4	ve	$<10^{-9}$	1/2	0	-1/2	0	9/36
	vu	$<1.7 \times 10^{-1}$					
	vt	$<1.55 \times 10$					

- Ref.—Wikipedia, the Free Encyclopedia (en.wikipedia.org/wiki/List_of_particles)

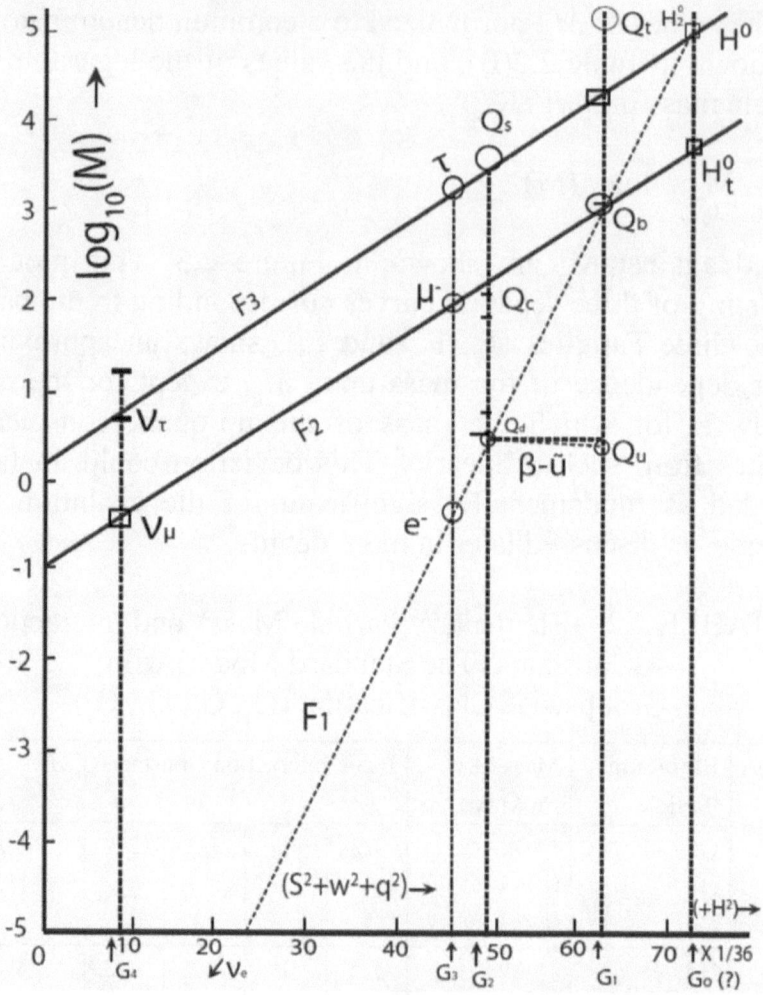

Figure 2.3. Plot of the particle mass $M_m = \log 10(M)$ against the value of M_0. The four 'Groups' (G_1, G_2, G_3, G_4) and the three Families (F_1, F_2, F_3) of the particles are marked. The extrapolation of the Standard Model elementary particles to include the Higgs bosons (H^0, H_1^0, H_2^0) is also shown. The departure of the Family 1 particle up-quark (Q_u) from linearity is marked by the β-decay (represented by the so-called β-Triangle.)

2.11. Linear Approximation for Mass-Interaction Relationship and Extension to Include the Higgs Bosons

2.11.1 The Predicted Higgs Bosons and the Corrected Neutrino Masses

The standard Model includes the 12 fermions (spin, s=1/2), but the results of the preceding Section could be extended to include the Higgs bosons—and these results also indicate the possibility of three Higgs bosons corresponding to the three Families, respectively. Even higher levels of extension (e.g., for gravitons with interaction constant = 2) may in principle be examined. Furthermore, perhaps better estimates of the three neutrinos could be obtained, again assuming a linear relation to hold down to the lower limits of the curves involved.

Assuming a linear relationship as shown in Figure 2.3 to be indeed valid for lower M_0 line-segment of Family 1 and for the entire ranges of Family 2 and Family 3, one may estimate, by extrapolation and interpolation, the approximate values of the masses of the neutrinos, at least in order of magnitude. Corrections could also be made in the mass-values of some of other particles. More importantly, the approximate mass(es) of the hypothetical Higgs particles could be predicted or theoretically estimated For the last item (Higgs boson), it is necessary to make an assumption for the value of the associated parameter of interaction, similar to the value of M_0 above. Assuming the values as follows:

$$S = 1; w = 0; q = 0; H = 1;$$

where we have arbitrarily introduced the 'Higgs interaction parameter' or 'Higgs charge,' H = 1 for the Higgs bosons, with the obvious assumption that H = 0 for all other elementary particles (fermions) of the Standard Model. Consequently, the sums of the squares of interaction charges for all the elementary particles discussed remain the same as before, except for the

Higgs bosons for which, we have the effective value of the interaction parameter (squared) given as

$$M_0 = S^2 + w^2 + q^2 + H^2 = (1)^2 + (0)^2 + (0)^2 + (1)^2 = 2 = 72/36$$

where the normalization by division by 36 has been introduced for convenience in graphical plot in order to include such a large range of variation of the mass-values. Also, the electric charge and the weak interaction parameter for the Higgs particle have been set to be equal to zero.

The vertical line at $M_0 = 2$ (= 72/36) intersects the three straight lines corresponding to F_1, F_2, F_3 (with the F_1-line extended linearly as shown in Fig. 2.3) at three mass-values, which could be taken as the *predicted Higgs masses*, denoted here as H_1^{0}, H^0, and H_2^{0}, respectively.

Thus, according to this model, we have a few 'corrected' or 'predicted' mass-values of the Standard Model particles, other mass-values being approximately the same as shown in Table 2.2. The results for these corrected or predicted masses are summarized below:

Particle	'Corrected' (C) or 'Predicted' (P) Approximate Mass-Values (M_a)		
		Log_{10} (M_a)(From Fig. 2.3)	M_a in MeV (or GeV)	(C/P)
Higgs:	H_2^{0}-(F_1)	5.35	223872.114 MeV	(P)
			(= 223.872 GeV)	(P)
	H^0-(F_3)	5.1	125892.541 MeV	(P)
			(= 125.892 GeV)	(P)
	H_1^{0}-(F_2)	3.75	5623.413 MeV	(P)
			(= 5.623 GeV)	(P)
Neutrinos	v_e	-9	10^{-9} MeV	(P)

	v_μ	-0.41	0.389 MeV	(P/C)
	v_t	0.74	5.495 MeV	(P/C)
Quark	Q_t	4.2	15848.9 MeV	(C)
			(= 15.849 GeV)	(C)

It is remarkable that the value of the H^0-mass 'predicted' (125.9 GeV), corresponding to the Family F_3, agrees well with the recently (July 4, 2012) 'tentatively discovered' mass of the Higgs boson (125.3 ± 0.4 GeV) at the CERN, by two experimental Groups, using the Large Hadron Collider (LHC) machine. The other two 'Higgs particles,' H_1^0 and H_2^0, with 'predicted' masses of 223.9 GeV (199.5 GeV if the y-value is read as 5.3 instead of 5.35) and 5.6 GeV, respectively, corresponding to F_1 and F_2, respectively, could be considered as additional Higgs under degeneracy (splitting) subject to the Higgs interaction.

Furthermore, these results, if experimentally confirmed, makes the electron-neutrinos (v_e) all the more unlikely candidate for the dark matter, as suggested by some. Indeed, the neutrino particles v_t and v_m corresponding to the 'inert' but heavier-mass Families F_2 and F_3, respectively, could be considered to be the more likely candidates for the dark matter, probably in the form of the weakly interacting massive particles (WIMP), as indicated before.

Barring the deviation for the up-Quark (Q_u), the three main line segments of Figure 2.3, corresponding to the three Families, F_1, F_2, F_3, respectively, of the elementary particles, can be generally represented by the equation of a straight line; so that the particle mass can be approximately specified by the simple formula:

$$Log_{10}(M_a) = m' M_0 + c'$$

$$= (S^2 + w^2 + q^2) \tan \xi + c' \, , \qquad (2.7a)$$

i.e., $$M_a = 10^{(m'M_0 + c')} \, , \qquad (2.7b)$$

where the suffix 'a' on the left denotes 'approximate' value, m' is the slope of the straight line in question, ξ is the angle the line segment in question makes with the horizontal axis, so that c' is the corresponding intercept with the vertical axis and $\xi = \tan^{-1}(m')$. From an examination of Figure 2.3, we obtain the approximate values of the angle ξ and of the vertical intercept c' as summarized in Table 2.3. Eqn. (2.7), together with Table 2.3, furnishes the desired, quaternion-based nonlinear (quadratic) model for the Unified Theory of Elementary Particles and Interactions (UTOEPI) postulated here.

TABLE 2.3—Values of the slope (ξ) and the vertical intercept (c') for a straight line representation or approximation of the mass-curves of Figure 2.3 (see Eqn. 2.7).

FAMILY	ξ (radians)	m'=tan(ξ)	c'
F_1	1.26	7.183	-9
F_2	0.56	2.375	-1.0
F_3	0.56	2.375	0.12

The line segment representation indicated above could be used to obtain a number of interrelations among the values of the masses of the particles involved. For example, by exploiting the geometrical properties of *similar triangles* (viz., '*the ratios of corresponding sides are equal*') obtained by the three straight lines corresponding to the three Families (F_1, F_2, F_3) in the graph of Figure 2.3 (out of which the latter two lines—for F_2 and F_3—are parallel themselves), and the mutually parallel vertical lines corresponding to the four Groups (G_1, G_2, G_3, G_4) shown in Table 2.2, one can derive a set of relations among the masses and parameter values of *different* elementary particles. Some examples follow to illustrate this point.

2.11.2. Example of Application of the Linear Relationship

Example—We denote a point location, and also the logarithm (with base 10) of the mass in MeV, of an elementary particle 'A' simply as A in the graph of Fig. 2.3, and a triangle with vertices at the location of the three particles A, B, and C (say) by the symbol ΔABC. Considering now $\Delta H_2^0 Q_b X$ and similar $\Delta H_2^0 Q_d X'$ and $DH_2^0 e^- X''$, where X, X', and X" (not shown in Fig. 2.3) are the points of intersection of the horizontal lines drawn through Q_b, Q_d and e^-, respectively, with the vertical line at $M_0 = 2$ (=72/36), we can write:

$$H_2^0 / (72\text{-}22.5) = H_2^0 X / Q_b X = H_2^0 X' / Q_d X' = H_2^0 X'' / e^- X''$$

where the first ratio pertains to the large triangle with vertex at H_2^0 and base in the M_0-axis (horizontal line at the bottom of Fig. 2.3). Substituting appropriate numerical values, the above equation becomes:

$$[5.3\text{-}(\text{-}5)] / (72\text{-}22.5) = (5.3\text{-}3.03) / (72\text{-}61) = (5.3\text{-}0.5) / (72\text{-}49) = [5.3\text{-}(\text{-}0.3)] / (72\text{-}45)$$

The values of the above four ratios are 0.208, 0.206, 0.209, and 0.207, demonstrating the equality of these ratios, as expected (difference in the third decimal place being due to error in reading the ordinate values.) Similar results can be obtained by considering the similar triangles $\Delta H_2^0 Q'_t Q_b$, $\Delta H_2^0 Q_s Q_d$, $\Delta H_2^0 \tau e^-$, $\Delta Q_b Q'_c Q_d$, $\Delta Q_b \mu\text{-}e\text{-}$, and $\Delta H_2^0 H_1^0 Q_b$; here the primes indicate the use of the 'corrected' numerical values of the particle mass, as discussed above. Other pairs and sets of similar triangles can also be utilized to correlate the particle-masses, to 'correct' the related numerical values, and even to 'predict' new particles, as exemplified by the H^0, H_1^0 and H_2^0. Thus the linear representation in the model presented here is of immense significance.

If the numerical values involved do not exactly satisfy such derived relations, the implication would obviously be either, or partially both, of the following conclusion(s): (a) The straight line representation or approximation for the curves in Figure 2.3 is less than satisfactory; (b) The measured value(s) of the mass of one (or more) of the particles involved is not accurately correct.

As demonstrated above, a straightforward generalization of the present model to include additional features, such as Higgs field, is readily achievable. Incorporation of the Higgs field has, as shown above, been accomplished by adding a term (H^2, say) to appropriately represent the field strength parameter on the right of Eqn. 2.5. This type of analysis could also be used to predict the value of the mass and the associated quantum properties (interaction parameters) of the Supersymmetric particles (sparticles) and other heavier particle. In reference to the above results (Fig. 2.3), this simply implies an extrapolation of the pertinent straight line (particularly for F_3, which corresponds to the heaviest particles) to an appropriately scaled higher value of the parameter M_0 (viz., 2 = 72/36, in Fig. 2.3).

2.12. Cosmological Evolution of Matter

The simple model presented above could also help one to understand the cosmological evolution process. This process could be conceived as consisting of a sequential introduction of the four basic interactions, one-by-one, in the sequential order or phases outlined below:

(1) *The 0-D Phase*—This refers to the pre-big bang situation, characterized by an absence of space-time and material objects, although one can still associate this phase as potentially endowed with an infinite amount of latent energy (and, hence, virtual mass) and energy-density.

(2) *The 1-2-3-D S (Spatial Space, r) Phase*—big bang creation of matter (quarks): onset of the spin of the pertinent elementary particles (quarks.)

(3) *The 4-D ST (r, t) Phase*—onset of the strong interaction: binding quarks to form neutrons;

(4) The 5-D ST Phase—onset of the weak interaction: neutron decay to yield protons and electrons (and anti-electron-neutrinos);

(5) The 6-D ST Phase—onset of nuclear fusion process and the electromagnetic interaction: providing bound states of protons and electrons to sequentially form Hydrogen, Helium and Lithium nuclei, and then simple atoms, under continued cooling of the background cosmic radiation, over cosmic time-scales ranging from the Planck time to hundreds of seconds and hundreds of thousands of years;

(6) Later Evolutionary Phases—onset of the gravitational interaction: attractive force between bodies with mass, and composed of atoms and molecules, to form larger bodies (stars, galaxies, galaxy-clusters, super-clusters) composed of atoms and molecules; and, subsequently, under favorable thermodynamic, chemical and biological conditions (e.g., in the interior of stars and on planets), of successively heavier nuclei and more complex inorganic and organic molecules; evolution and proliferation of life on the Earth; and so on.

The above description is focused on the Family 1 elementary particles. Family 2 and Family 3 particles do not undergo the weak interaction-induced nuclear decay process to continue the above evolutionary pattern; although, being more massive and assumed here to act as the dark matter, they presumably play crucially important roles in the formation, evolution, and stabilization of stars and galaxies under the gravitational interaction, as mentioned before.

Needless to say, the evolutionary process described above occurs on a cosmic scale of space and time, of the energy—and temperature—levels, and of the system-size—the changes involving thousands and millions of orders of magnitude, the initial changes involving space-time on the Planck scale. In general, each phase of the evolution and each class of interaction could be characterized by a specific interaction potential function; and the phase-changes could be assumed to take place subject to certain critical values of the system size, temperature—and energy (or mass)-levels, and other basic parameters. The phase transformations could therefore be even describable in terms of these parameters in the general framework of the 6-D space-time-based cosmic thermodynamics. It is customary in thermodynamics to use the concept of the 6-dimensional 'phase-space' constituted of the dimensions of space and '3-dimensions' of the momentum vector. In this 6-D 'phase-space,' the velocity is a vector with components along the three coordinate axes (of space) delineating the rate of change of the position with respect to a single time-component (treated as a scalar.) In the present model, however, time itself is taken as a 3-D vector, so the velocity (v), and hence also the momentum (p), could be treated as tensor of order 2, with the space-time 'curvature' being represented by a metric or a 3x3 matrix (e.g., $v_{\mu v} = \partial X_{\mu} / \partial T_{v}$). Of course, the concept of the thermodynamic phase-space is relevant only for a large assembly of particles interacting in a manner describable in statistical terms.

Inherent 5-fold symmetry in the spherically symmetric systems involved preserves the basic interaction parameters, i.e., "charges" ascribed to an elementary particle, throughout the various phase changes. The evolution from the Family 3 to Family 2, and then to Family 1 elementary particles, if it indeed took place in that order, resembles the biological evolution process on the Earth, since Family 1 particles and processes are ultimately responsible for the chemical and biological evolution in the universe, as mentioned earlier in this book and succinctly further explained below.

2.13. The Especial Significance of the β-Decay for Evolution

2.13.1. General

Of critical importance in connection with the cosmological (and, by extrapolation, with the biological) evolution process is the nuclear β-Decay. We review below this well-known process from the stand-point of its impact on the classification of the elementary particles, on the interconnection and unification of the interactions, and on the related evolutionary phenomena in the universe in general and on the Earth in particular.

We readily notice an important feature of Figure 2.3, namely, a downward turn of the Family 1 'curve' (line segment) in the extreme right corresponding to the high M_0-value region. This feature obviously implies a decrease in the value of the mass of the up-quark in comparison to that of the down-quark. This feature is conspicuous particularly in contrast with the persisting linear rise (without a similar downward turn) of the analogous values in the cases of Family 2 and Family 3, respectively. This feature [viz. the inequality $M(Q_d) > M(Q_u)$] is in fact the basic factor permitting, as well as explaining, the weak interaction-induced nuclear β-decay; i.e., a neutron (n) decaying into a proton (p^+), an electron (e^-) and an anti-electron-neutrino (v_e^*). This nuclear reaction can be also interpreted as the weak interaction-induced decay of the down-quark (Q_d) into an up-quark (Q_u), an electron (e^-) and an anti-electron-neutrino (v_e^*). Conservation of mass (M) and energy would require that we must have the condition $M(Q_d) > M(Q_u)$; because the condition $M(Q_d) < M(Q_u)$ would clearly disallow the β-decay of the neutron and, hence, all nuclear decay processes. As a matter of fact, a reverse variation pattern for the Family 2 and Family 3 particles *does* disallow an analogous decay or transformation process or reaction. The overall conclusions from these observations are as follows.

2.13.2. Anthropic Principles

Family 1 quarks, normally bound under the strong interaction, when subject to the weak interaction under lower energy—and temperature-values, undergo the nuclear β-decay of the neutral neutron, yielding the (electrically) oppositely charged particles—the proton and the electron. Under the electromagnetic (EM) interaction—the next hierarchical interactions from the stand-points of interaction strength, energy-levels, cosmic background radiation temperature, particle-size, etc.—the proton and the electron combine to form the Hydrogen atom. Other light nuclei (Deuterons, Helium, etc.), formed under nuclear fusion also form corresponding atoms under the EM interaction. Through a repetitive process of generation of successively heavier nuclei in stellar interiors, stellar and galactic explosions, and of recombination of positively charged nuclei and electrons, successively heavier atoms and molecules are formed. This marks an end of the 'dark era' since these neutral atoms (and molecules) allow radiation to pass through cosmic distances of billions of light-years, unaltered, to make the optically emitting part of the universe visible. More and more complex (organic and inorganic) molecules, and various complex organisms including biological cells, DNA, and so on, are formed, producing myriads of inorganic substances as well as countless living species, including man, through the familiar processes of Darwinian Evolution. Thus the big bang through the Darwinian evolution could be considered to be different parts of similar evolutionary processes; like different 'colors' of the same continuous cosmic 'spectrum,' with probabilistic transformations of species from one form to the next, from the quark to the DNA.

This series of events has a direct bearing on the so-called Strong Anthropic Principle[3] (SAP.) The so-called Weak Anthropic Principle (WAP) assumes that an infinite number of universes might exist, of which our universe is one that supports life as we know it. In contrast, one may postulate a Medium-, or Natural-, or Life-based Anthropic Principle (abbreviated as MAP, NAP or LAP, respectively), stating that "Life as we know it originates, evolves and survives in a self-consistent manner only in a universe of which we are a product and which, therefore, we can know and naturally conclude that its conditions are just right for Life."

2.13.3. The Higgs Field

Possibility might well exist of higher mass-state(s), in the form of a more massive Family (Family 4?) of elementary particles or, most probably, of a higher M_0 value corresponding to the inclusion of additional interaction field such as the Higgs field. Supposedly such a massive particle (or equivalent field) is normally not revealed within the scale of the values of the variables spanned by those of the Standard Model, and could presumably become realizable at sufficiently higher energy range, as is commonly believed. Properties of such a massive particle or field may be obtained by extrapolating the Standard Model. Such an extension will naturally lead to 'new physics,' thereby probably also *'justifying'* the existence of the Family 2 and Family 3 elementary particles in a more concrete way. Future experimental observations from the Large Hadron Collider (LHC) or even more powerful machine are likely, and anticipated, to help resolve this question. Such observations would naturally also throw light on the features inherent in such additional particle(s) and field(s), and hence on the need for an appropriate extension or generalization of the Standard Model and of the UTOEPI proposed here. Clearly, a generalization of this UTOEPI in order to reflect such elements of 'new physics' is relatively straightforward, as mentioned earlier in this book.

2.14. Bifurcation Diagram Representation

The Families of elementary particles and their interactions could be also represented in the form of a simple bifurcation diagram, as illustrated in Figure 2.4 (a, b).

In the *chaos theory* pertaining to *nonlinear feedback effect* in a *complex system*, *bifurcation* is commonly associated with *order-disorder* or *phase-transition phenomena* and is usually associated with the *self-similarity* property of the system. Here, *the evolution of the four Groups of elementary particles, G_1, G_2, G_3, G_4, is apparently a systematic phase-transition due to onset of the hierarchy of interactions—strong, weak, EM and gravity, in that order—with decreasing cosmic temperature—and energy-level following the big bang*, as mentioned above. Also the three Families of particles, F_1, F_2, F_3, respectively, exhibit signs of self-similarity and evolutionary discrete transitions in the interaction parameter values. For instance, in going from the quarks to the leptons, the strong interaction charge (S) as well as the weak interaction charge (w) decreases by one unit (from $S = 1$, $w = 1/2$ to $S = 0$, $w = -1/2$). Similarly, the EM charge (q) decreases by one unit in going from up-quark to the down-quark (from $q = 2/3$ to $q = -1/3$), and increases by one unit in going from the electron to the neutrino (from $q = -1$ to $q = 0$). These aspects of the system of elementary particles and conserved parameter values under the basic four interactions, viewed as manifestations of a universal evolutionary process (analogous to the genetic transformations and preservation of the DNA structure in living species) need further investigations. Naturally, such investigations bear the promise of outcomes of fundamental importance regarding the very nature of Space and Time from the very small to the very large scale, as well as the universe as a whole.

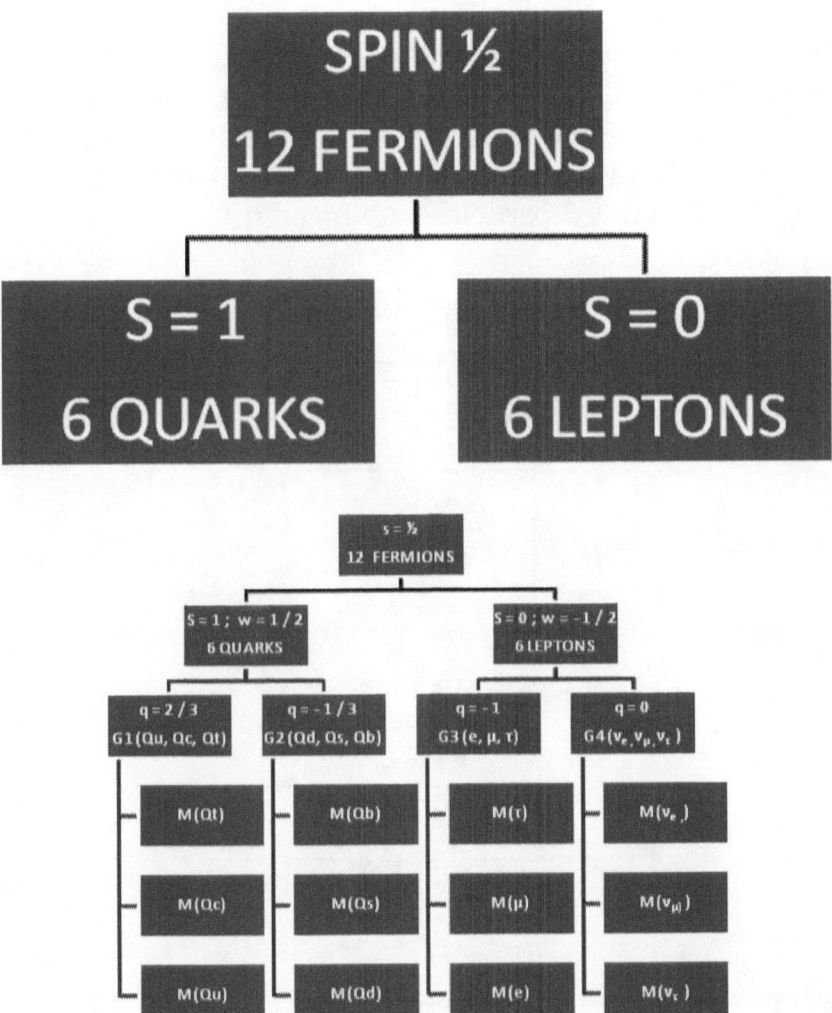

Figure 2.4   (A).—Bifurcation Diagram for the Standard
              Model Elementary Particles and Interactions:

(a) Upper Diagram—Basic pattern of bifurcation
              (strong interaction);
(b) Lower Diagram—Bifurcation under successive
     introduction of strong, weak, EM and
     gravitational interactions.

An alternative representation in terms of a simple line-diagram is presented in Figure 2.4(b) with labels for the particles in terms of the 6-D ST coordinates (x_1, x_2, x_3 and t_1, t_2, t_3) also provided at the bottom of Figure 2.4 (b), with a tentative association of Family 1 particles with the positive arrow of time (t_1) (and of Family 2 and Family 3 particles with t_2 and t_3.)

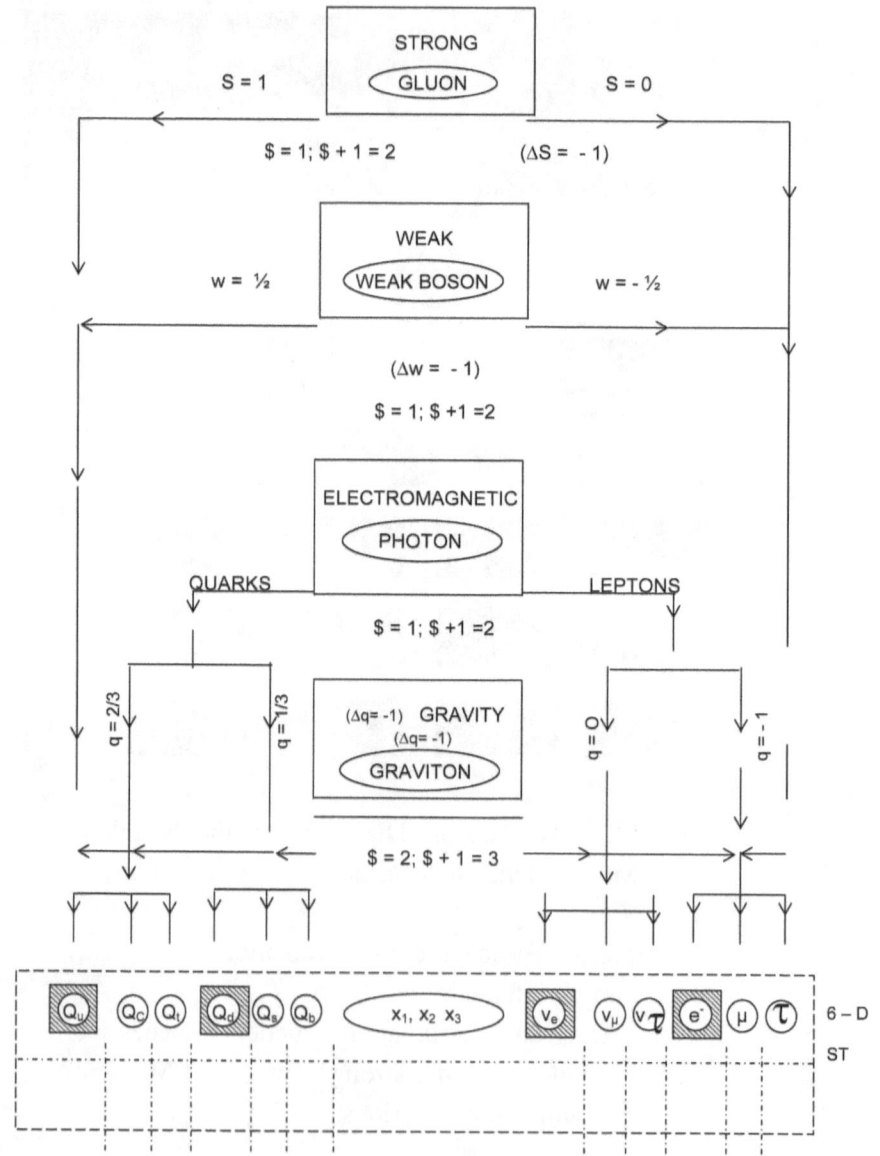

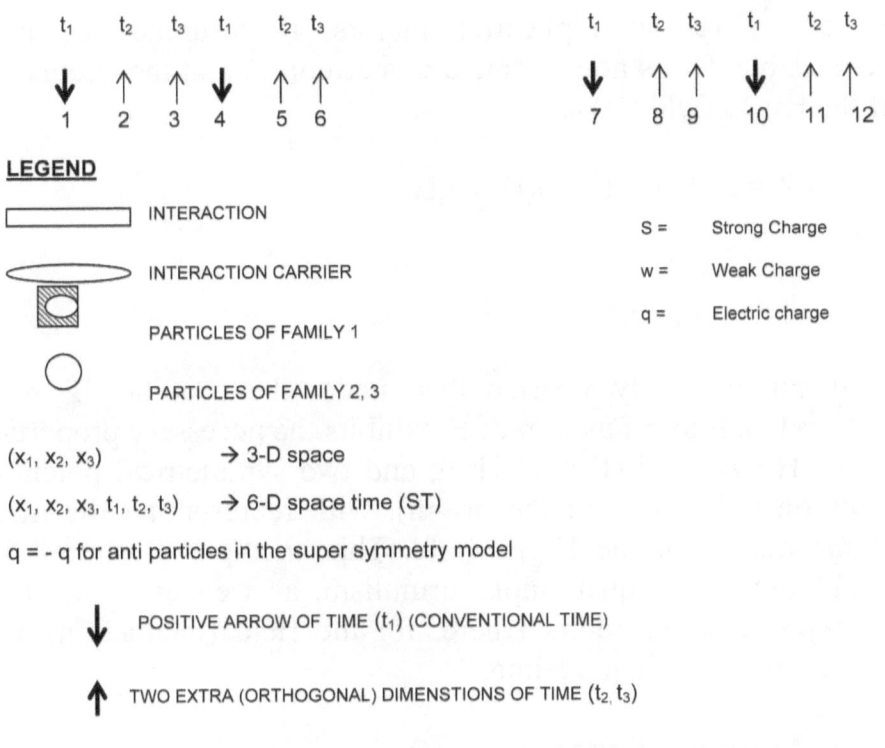

Figure 2.4(b). Diagrametical Representation of Basic Interaction and Elementary Particles (QUARKS, LEPTONS; $ = SPIN = ½; 2$ + 1 = 2)

2.15. Modeling The Higgs Field

In contrast to the reduction into the 2-D hyper-plane, in the extended relativity theory (ERT) propounded in the present paper, the mathematical properties of the quaternion may be exploited with advantage to derive properties of the system of fields and elementary particles in the (r, icx_4, icx_5, icx_6) hyperspace, without imposing any restriction on the relative velocity vector (see Eqn. 1.3 a,b). As an example, a simple formulation of the Higgs field (if it exists), can be obtained as follows.

In Eqn. (2.4.e,f) above, let the coefficient a_o be equal to the radial space coordinate r; and let the coefficients of a_1, a_2, a_3 be the (imaginary) time coordinates ict_1, jct_2, and kct_3, respectively.

Since any real multiplicative factors, if desirable, are also permissible, let us now define a quaternion Z, and the energy, E, of the Higgs field (H) as

$$Z = a_0 H^2 + a_1 H + a_2 H + a_3 H$$

$$E = -Sm_{=1}^{3} a_m^2 H^2 + a_0^2 H^4 \quad ; \tag{2.8}$$

It can be easily verified that, under the condition $r^2 = c^2$ $(t_1^2 + t_2^2 + t_3^2)$, E as a function of H exhibits the necessary properties of the Higgs field (E=0 at H=0; and two symmetrical potential wells on either side of the origin), with representing the mass of the quanta of the Higgs field. This example illustrates the usefulness of the quaternion formulism, as well as of the 6-D ST representation adopted here, for the field (interaction) and elementary particle modeling.

2.16. Further Discussions

2.16.1. General Background: recapitulation

The well-known formulation of Einstein's Special Relativity Theory (SRT) and General Relativity Theory (GRT) is based on the premises of:

a) Lorentz transformation of the Cartesian coordinates of a frame moving with respect to a reference frame with a relative velocity (v), usually taken along the common x-axis (say);

b) A four-dimensional space-time (4-D ST) continuum, with three spatial orthogonal coordinates (x, y, z) and a single time coordinate (ict), assumed to be orthogonal to this three-dimensional space (3-D S), where $i = \sqrt{(-1)}$, and c is the velocity of light (see Eqn. 2.9);

c) Intrinsic curvature in the 4-D ST due to the presence of mass (M), determining the trajectory of particles as the geodesics in the curved 4-D ST geometry.

Here we have presented arguments to highlight the need to extend the concept of 4-D ST to a new paradigm of the six-dimensional space-time (6-D ST) continuum by introducing a three-dimensional time (3-D T) space orthogonal to the 3-D S, with the three time coordinates also forming a right-hand orthogonal Cartesian reference frame similar, and orthogonal, to the 3-D S. Apart from the conventional time (associated with the "positive arrow" of time of increasing entropy and of common physical phenomena and every-day experience), the additional two orthogonal time dimensions may be regarded analogous to the familiar Kaluza-Klein ("curled-up") variables playing a significant role only in ultra-high energy and Planck scale space-time range pertinent only for sub-nuclear particles created in the big bang and in high-energy particle accelerator, for example, and in cosmological processes.

$$(x_1, x_2, x_3, x_4, x_5, x_6) \rightarrow (x_1, x_2, x_3, x_4) = (x, y, z, ict) \ldots \rightarrow (r, t_r) \qquad (2.9)$$

$$<\ldots 6\text{-D ST} \ldots > <\ldots 4\text{-D ST} \ldots> <\ldots 2\text{-D ST} \ldots>$$

It must be recognized that the additional time-dimensions in the 6-D ST are simply measure of time for motion and variations in the transverse directions with respect to the direction of observation; and, as such, there is no mysterious connotations associated with them.

The well-known fact that the strong interaction between two quarks increases with increasing separation between them naturally allows a simple modeling of the same in terms of a "spring"-like force or potential; and the related formulation of the strong interaction may be referred to as the 'spring theory.' The electromagnetic and gravitational interaction potentials vary in inverse proportion to the radial separation. Thus the radial separation between two points, specifying the locations of the two interacting systems (particles or cosmological bodies, for example) with respect to the observer is of prime interest in most cases.

Of fundamental and critical importance in interpreting observed experimental data in high-energy physics and cosmology is the apparent separation which varies with the relative position of the observer. This is because, for two or more particles or bodies, the system behavior is governed by the relevant force-fields F(r) or, equivalently, by the potential functions V(r), [F(r) = $\partial V(r) / \partial r$] which might be dynamically changing in the 3-D S and the 3-D T spaces, i.e., in the 6-D ST hyperspace, due to their critical dependence on the instantaneous mutual separation, r, especially for a spherically symmetric system. However, the value of this separation is critically dependent on their relative location with respect to a particular observer, as shown by the referred set of equations. This implies that, apart from the well known relativistic length-contraction and time-dilation effects (operative for a 'radially situated' observer), as predicted by SRT, there exists apparent variation of the pertinent space-time separation between the two interacting bodies, as interpreted by any observer, with explicit dependence on the bodies' coordinates with respect to the observer (see Sections 2.19 and 2.20 below for details.) A change of location of the observer with respect to the interacting system could thus produce an illusory effect indicating a different behavior pattern by the system. This happens because the angular coordinates change with a change in the location of the observer with respect to the two points interacting mutually, leading to a different value of the relative separation between them; and hence the values of the potential function and the force-field change accordingly, even though no such change should in principle occur in the system in absence of any *external* force.

In particular, *two observers with distinct locations with respect to the system would interpret the data from the same event differently.* We term this effect "Observer-Location Dependent Effect **(OLDE)**." Clearly this effect can be quantitatively determined and compared with the value predicted from a theoretical analysis. In this connection it must be noted that the 'observer' could be a particle detector in a high-energy accelerator system, or a terrestrial or space-based telescope; and its location must be accounted for in interpreting the observed data. In case of a cosmological event, the variation in the location of the observer can obviously be considered relatively negligible due to the 'astronomical' distances involved; whereas it may play a critical role in the case of particle-accelerator related observations.

2.16.2. Cosmological Problems

Some of the prominent unresolved questions of cosmology and particle physics are as follows:

(a) What is the exact nature of the "positive arrow" of Time?
(b) What constitutes the dark matter?
(c) What is the origin of the dark energy?
(d) What is the mechanism of the cosmological inflation?
(e) How do we explain the entanglement phenomena?
(f) Why do we not see anti-matter (created at the big bang in equal amount as matter)?
(g) Why do we have just the elementary particles and interactions we do, and not others?
(h) Why do the Family 2 and Family 3 elementary particles exist?

The above 'problems' are considered below qualitatively in the light of the 6-D ST concept to illustrate, explain, and explore how this concept could help resolve them.

The problems of the positive arrow of time, the dark matter, and the dark energy have already been considered before. It is also interesting to further note the following additional factors.

2.16.3. The Dark Matter

The above result also indicates that the mysterious dark matter in the universe can, in all likelihood, be identified with the large amount of inert matter of Family 2 and Family 3 elementary particles around galaxies and galaxy clusters. Such inert products of the big bang, practically unchanged throughout the age of the universe and pervading the universal 6-D space-time continuum, might in fact have played a significant role in the formation and evolution of galaxies and galaxy-clusters composed of visible matter, by virtue of their large gravitational interaction effects. Thus the hypothesis of the 6-D ST (i.e., 3-D T), together with the principles of DEM and FEM (see Sec. 2.1), may be relevant in answering a number of questions in elementary particle physics and cosmology; exemplified by the following questions and answers (Q&A):

(1) Q—What is the relevance and role of the seemingly inert Family 2 (F_2) and Family 3 (F_3) elementary particles?
A—These relatively heavy particles likely play a significant role in constituting the so-called 'dark matter' which, in turn, probably have influenced the formation of early galaxies and galaxy-clusters out of the seemingly uniform distribution of visible matter in the early phase of the universe by virtue of their large masses and, hence, large gravitational forces. The existence of two types of 'dark matter', axions and neutralinos, likely correspond to two such Families, F_3 and F_2, respectively, of the elementary particles, in view of their relative masses (energy-values) and the estimated time-periods elapsed after the big bang for their synthesis.

(2) Q—Why is the dark matter 'dark' (invisible)?
A—The pertinent elementary particles do not evolve to form atoms and molecules that emit radiation for an observer to "see" (optically detect) them.

(3) Q—Why don't the Family 2 and Family 3 particles evolve to form atoms and molecules?
A—These particles do not undergo the nucleonic decay process (see Section 2.13.)

(4) Q—Why do the Family 2 and Family 3 particles behave in a radically different manner compared to the Family 1 particles?

A—Family 1 particles pertain to the "real" ("positive arrow" of) time to which all commonly observable matter and the related biological evolutionary products—including ourselves—belong; whereas the particles of the Family 2 and Family 3 pertain to virtual or "imaginary" transverse time dimensions, hidden from us and from any associated direct observations.

(5) Q—How did the primordial, uniformly distributed, matter from the big bang become organized into clumps forming galaxies?

A—Through the large gravitational interaction of the original dark matter of the Family 2 and Family 3 particles.

2.16.4. The Dark Energy

The question of the dark energy could be theoretically examined in terms of quantum uncertainty-induced generation of energy and polarization of the same in the 'angular' time dimensions in the 6-D ST. This can be seen by writing the conventional quantum uncertainty relations for energy (E) and time (t) and for the position (x) and momentum (p), as applied in the 6-D ST:

$$\Delta E_j \Delta t_j \sim \hbar \sim \Delta x_j \Delta p_j \,, \; (j = 1, 2, 3)$$

i.e.,
$$\Delta E_j \sim \frac{\hbar}{\Delta t_j}$$

and
$$\Delta x_j \sim \frac{\Delta E_j \Delta t_j}{\Delta p_j} \sim c \Delta t_j \,,$$

where $\hbar$ is the Planck constant ($\hbar \approx 1.05$ x 10^{-27} erg s.); and we have used the relation $\Delta E_j = c(\Delta p_j)$ valid for radiation and also for ultra-energetic particles moving with velocities close to the velocity of light, since then $E \approx mc^2$ and $p \approx mc$, so $E \approx pc$.

Considering the 3-D T time components t_j (j = 1,2,3), The above uncertainty relations apply for conventional time (j = 1) as well as for the orthogonal time components (j = 2, 3), with correspondingly additional energy 'components.' Clearly, miniscule uncertainties in orthogonal time components (comparable to the Planck time) could be potential sources of 'dark energy' of huge proportions in the cosmic domain.

The 'dark' (invisible) characteristics of these energy components presumably refer to their being polarized in the transverse (longitudinal and/or latitudinal) directions of the three-dimensional time-space, i.e., perpendicular to the conventional time (t_r) of experimental observations. Orthogonal polarization of radiation would normally disallow observation or detection of such radiation. This cross-polarization-blockage property, well-known in the classical electromagnetic theory, is now-a-days widely and routinely exploited in scientific and commercial satellite telecommunications for achieving a higher degree of efficiency in the utilization of the limited resource of the available radiation bandwidth, through the so-called 'frequency (spectrum)-reuse' based on 'cross-polarization discrimination' or 'polarization isolation.' This refers to the technique of repeated use of the same spectral bandwidth, but with the signal transmitted or received through emissions which bear spatially orthogonal polarizations. If it were feasible to 'rotate the plane of polarization by 90°' to coincide with the plane of the 'normal' time, such 'dark energy radiation' should in principle become observable.

2.16.5. Cosmological Inflation

The question of the cosmological inflation process just following the big bang could be theoretically examined in terms of an effective 'conversion of the time dimension into the space dimension' in the 6-D ST. Specifically, in the spherical polar coordinate system, with or without any amount of uncertainty in Δt_r, there might exist, now or at an early phase of the universe including immediately after the big bang, arbitrarily small Δt_θ and/or Δt_ϕ, thereby leading to the possibility of cosmological generation of very large amount of the dark energy as well as sudden *spatial* expansion. For instance, for $\Delta t_j \sim 10^{-43}$ sec. (Planck time), since $\hbar$ is of the order of 10^{-27}, a 10^{16} order of magnitude generation and/or enhancement in energy, and also a spatial expansion over a Planck distance, 10^{-33} cm, could be expected. This type of phenomena occurring repetitively could, therefore, possibly also form, or contribute to, the explanation of the cosmological inflation processes supposed to have immediately followed the big bang.

2.16.6. The Entanglement Phenomena

The entanglement phenomena refer to a synchronous or similar behavior of two particles too far apart to have a causal connection or communication between them. In essence, this implies that the spatial distance between the two particles is larger than that for which they could interact even via the fastest signal (traveling at the speed of light.) However, even if the distance involved be larger than ct_r, t_r being the radial time component of the time-space vector ($\underline{t}$) in the 3-D T, the angular separations of particles within the system of interest could be arbitrarily small in comparison to the component(s) $c\Delta t_\theta$ and/or $c\Delta t_\phi$, thereby permitting a connection and causal communication between the particles involved. Thus the hypothesis of a 3-D T could resolve the mystery of the entanglement phenomena.

2.16.7. The Anti-Matter Dilemma

A question often asked in cosmology is: Where have the particles of anti-matter gone [15]? A plausible explanation for the vanishing anti-matter in the context of 6-D ST is as follows.

Supposing that equal numbers of particles and anti-particles were originally created by the quantum vacuum fluctuation process that led to the big bang, the anti-particles would follow opposite trajectories in the 6-D ST in contrast to the particles due to their opposite electric charges. Thus we can suppose that as soon as the particle and anti-particle pair is created, the two members are whisked away into oppositely directed or segregated segments of the space-time continua in the $(x_1,x_2,x_3,x_4,x_5,x_6)$-space. An inversion of the spatial coordinates alone is not sufficient to guarantee a permanent separation of the populations of the two twins of the (particle and anti-particle) pair; a time-inversion would, however, make this separation complete.

It is possible in principle to also imagine a negative sign for the square-roots involve. As mentioned above, while we commonly assume only the positive sign for the time vector corresponding to the "positive-arrow" of time in which matter exists, interacts and evolves, the negative sign of the time vector is theoretically permissible and could be associated with the anti-matter. In effect, one can assume that all the particles and matter (including ourselves) move and interact only in the "positive-arrow" direction of time (with increasing entropy), whereas the anti-particles move and interact only in the "negative-arrow" direction of time. Due to this intrinsic separation in the space-time, the anti-matter is not accessible to our observation. Also, this intrinsic separation prevents an automatic annihilation of the two types of universes (one made of matter and the other of anti-matter.) Creation and annihilation of particle-anti-particle pairs in laboratory (particle accelerator) conditions is apparently possible since much lower energy and temperature ranges than those present at the big bang stage (Planck-scale) are involved.

2.17. Successive Introduction of the Higher Dimensions

In the following sub-sections, we first illustrate the effect of hierarchically introducing various basic interactions one-by-one.

The resulting degeneracy is simulated by the process of successive introduction of higher dimensions through an initial *translation* and subsequent axial *rotations*, starting from the zero or null-dimension (0-D). This process is schematically illustrated with the use of simple geometrical diagrams for clarity (also see Fig. 2.2).

2.17.1. Zero-Dimensional Space-Time (0-D ST): POINT

In the case of 0-D ST (*'POINT'*), the existence of elementary particles or interaction becomes a moot question, since, by definition, 0-D corresponds to an absolute 'nothingness,' geometrically as well as physically. This may be regarded as the situation *"prior to the big bang"*—'a non-existence of space, time or matter.' Figure 2.5(a) symbolically represents a 'Point,' marked by the *origin* 'O.'

⌷

O

Figure 2.5(a):   The 0-D case ['POINT']: Null space-time or
material content

2.17.2. One-Dimensional Space-Time (1-D ST): STRAIGHT LINE

The 1-D ST (*'STRAIGHT LINE'*) is the result, so to say, of a simple *rectilinear translation* of the reference 'Point' O. The 'Straight-Line' resulting from the translation of the 'Point' can be taken to define the coordinate axis, x_1, for representing the spin (s) of the elementary particles. The conserved spin values [½ for the fermions; 0, 1 or 2 for the bosons—gluons, photons, weak bosons ($W^\pm, Z$), and gravitons, respectively] is specified by the length of the line segment from the reference point (origin) O, as schematically shown in Figure 2.5 (b). Note that all 12 Standard Model elementary particles are clustered at a single point (marked X in Figure 2.5 (b) of the line, corresponding to the (s = ½) value for the fermions. The spin s=1 point could also be assignable to the 'sparticles' if so desired, so a potential generalization of the present model to include supersymmetry is obvious. Note also that the length of the line along the x_1-axis is to be taken as a *real* variable; i.e., x_1 is regarded as the *'real'* axis.

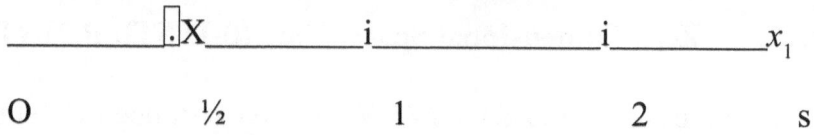

Figure 2.5(b): The 1-D case [STRAIGHT LINE]: Introduction of spin [The set of the 12 fermions of the Standard Model is clustered at the point marked X corresponding to spin s=1/2.

2.17.3. Two-Dimensional Space-Time (2-D ST): PLANE

Now consider a rotation of a straight line coincident with the x_1-axis of Figure 2(b), about an arbitrary axis of rotation, by $\frac{\pi}{2}$, taken as positive for anti-clockwise rotation, to yield the orthogonal axis x_2, as schematically illustrated in Figure 2.5(c). This process of *'ROTATION'* of a Straight-Line yields a 2-D *'PLANE'* defined by the mutually orthogonal axes (x_1, x_2). Let us formally identify the vertical axis with the geometric axis for the strong charge (S).

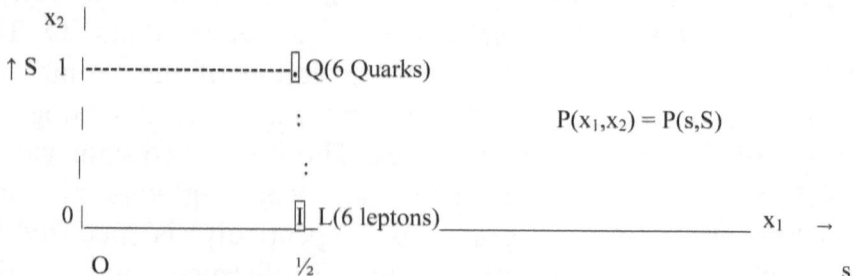

Figure 2.5(c).The 2-D case [PLANE]: Introduction of the strong interaction [Geometrical representation

of the spectral splitting of the 12 fermion particle set of Fig. 2(b) into two sets of particles, viz. 6 leptons (s = ½, S=0) and 6 quarks (s = ½, S= 1).]

This process of ROTATION of a Straight-Line can be utilized to distinguish two points in the (x_1, x_2)-plane, i.e., the (s, S)-plane; viz., L(1/2,0), corresponding to the 6 leptons (electron, muon, tau-particle, and associated neutrinos); and $Q(1/2,1)$, corresponding to the 6 quarks. In other words, the introduction of the strong interaction, here geometrically correlated to going into a higher (2-D) dimension through the physical process of rotation, splits the population of the 12 elementary particles into two distinct groups, leptons and quarks, in a manner analogous to the degeneracy of a hydrogen atom energy-level undergoing splitting into a hyperfine structure when subjected to a magnetic field (the Zeeman effect), as mentioned before.

We now consider the cases of still higher dimensions (n-D ST, n = 3,4,5,6) by introducing additional dimensions one-by-one. Let us skip the intermediate steps for brevity and examine the final configuration of the 6-D ST.

2.17.4. Higher-Dimensional Cases: Space-time up to 6-D ST Hyperspace

Continuing introduction of additional dimensions, we simulate the generation of the 3-D, 4-D, 5-D and 6-D space-time, in step-wise fashion, introducing one extra dimension at each step. We confine our attention to spherically symmetric systems and, hence, limit our consideration to the case of a 5-sphere and associated conserved parameters. For a graphical depiction, the radial coordinate, r, in the 3-D S space (with implicit symmetry in the spatial angular coordinates and thereby corresponding to the conserved spin value, s) can be omitted here for simplicity. Now let us re-label the $\hat{x}_2$-axis in Figure 2.5(c), taken as the fourth ('*imaginary*') dimension to correspond to the polar ("positive-arrow") time coordinate, t_r, at the point (r, θ, ϕ) of the 3-D S (*spatial* space with real coordinates) of the 6-D ST, and associate it with the conserved value of the strong charge (S). Further, through a rotation by $\pi/2$ radians of a line coincident with this (t_r) axis, we introduce the next higher (i.e., the 5th) orthogonal ('*imaginary*') dimension corresponding to the azimuth time vector for t_θ in the 6-D ST, and associate it with the weak charge (w). Similarly, the next and the last higher (6th) dimension is introduced by a final rotation of the reference axis by $\pi/2$ radians, and formally associate it with the conserved EM charge (q) value. This completes the four possible orthogonal rotations in the 6-D ST spanned by the coordinate variables, finally returning to original (real) dimension corresponding to r. At each step, the additional dimension or orthogonal axis is introduced following the conventional right-hand rule of a Cartesian coordinate system.

At each rotation, i.e., introduction of the next dimension and the associated interaction and its conserved parameter, each group of the elementary particles becomes regrouped into two smaller subgroups, since the locations of the group members are shifted ("split") in the space of the conserved parameters (s, S, w, q, M), originally reflecting the 5-fold symmetry in the 6-D ST. This "splitting," tantamount to a successive set of symmetry-breaking or degeneracy effect, occurs in a manner analogous to the transition from the 1-D case to the 2-D case discussed above (Fig. 2.5b to Fig, 2.5c.) Thus, at each stage, the redistribution of the relevant group of the elementary particles in the 5-dimensional space of the parameters (s, S, w, q, M) arises from the introduction of the appropriate 'generator' creating the next higher dimension in the quaternion hyperspace. This process is continued up to the inclusion of the five independent symmetrical dimensions of the 5-sphere. In fact, we can associate each particle with a hyper-plane consisting of two orthogonal dimensions. The total number of elementary particles is, therefore, equal to the number of separate hyper-planes one can construct out of a total of 5 dimensions, which equals the value of the binomial combination $^5C_2 = 10$, in agreement with the possible number of elementary particles derived earlier, subject to the assumption of the neutrino-mixing related reduction in the number of the elementary particles as hypothesized in Section 2.3.1.

2.18. The Quaternion Representation as Rotation in Higher Dimensions

To derive the desired relations among the interaction parameters taken as the basic properties of each elementary particle, recall that successive rotations in the 6-D ST, i.e., in the 4-D hyperspace $(r, ict_r, jct_\theta, kct_\phi)$, are mathematically represented by the generators (i, j, k) of a quaternion obeying the non-commutative algebra specified in Eqns. 2.4(a-e). The non-commutative properties of the quaternion reflect the analogous characteristics of the rotation operation: A rotation X followed by another rotation Y is *not* equal to the rotation Y followed by the rotation X (XY $\neq$ YX). The reality of physical rotations associated with the quaternion generators (i, j, k) as described above can be further illustrated with the help of an elementary example comprising repetition of a rotation using only the first generator i. This corresponds to rotation in the (2-D) plane twice, each time by $\frac{\pi}{2}$ radians and, hence, by a total of $2 \cdot \frac{\pi}{2} = \pi$ radians, leading to a reversal of sign or, equivalently, multiplication by (-1), as indicated in Figure 2.5(d).

$$\begin{array}{c} \text{I} \\ \frac{\pi}{2} \leftarrow | \leftarrow \frac{\pi}{2} \end{array}$$

-1_____ ✓_____|_____↖_____1

Figure 2.5(d) Double rotation, each by $\frac{\pi}{2}$ radians (i.e., by a total of π radians).

Mathematically, this repeated rotation is symbolically expressed as $i^2 = ii = e^{1\frac{\pi}{2}} \cdot e^{1\frac{\pi}{2}} = e^{1\pi} = -1$. Such a description of rotation in terms of the quaternion generators also similarly applies for j^2 and k^2. Thus, the defining algebraic relations of the generators (i, j, k) of the quaternion Z (Eqns. 2.4a-2.4f) are exactly consistent with the rules of rotation of the coordinate frames or axes in the physical space-time. This proves that the quaternion generators indeed represent a physical rotation of the coordinate axes in the 6-D hyperspace.

The final grouping of the elementary particles of the Standard Model resulting from the 'spectral splitting' in the 3-D orthogonal space of the conserved interaction parameters (S, w, q) is schematically depicted in Figure 2.2. This splitting process in fact leads to four distinct 'Groups' denoted here as G_1, G_2, G_3, and G_4, respectively. Each Group comprises 3 particles having identical parameter values. This feature is exploited below toward further quantitative analysis, and can be said to form the basis of the simple model (UTOEPI) presented here. Note that the final rotation brings the reference axis to a position that corresponds to the initial real axis along which the last interaction in this hierarchy, namely, gravity, affects the redistribution of the particles in the hyperspace. Introduction of this last interaction—gravity—is seen to introduce splitting of one particle state with identical value of s, S, w, and q into three degenerate members. This is consistent with the basic characteristic of the interaction-carrier, viz. the graviton, which possesses a spin-value of 2 and hence possesses a degeneracy of 3. The three degenerate mass-values correspond to the elementary particles of the three Families, F_1, F_2, and F_3, respectively, of the Standard Model.

2.19. Spherical Polar Coordinates in the 6-D ST

In case of an intrinsic spherical symmetry of a system, it is obviously more convenient to use the orthogonal spherical polar coordinates $\underline{r}$ (r, θ, ϕ) in the 3-D S (*spatial*-space) and the corresponding orthogonal spherical polar coordinates ic $\underline{t}$ (ict_r, ict_θ, ict_ϕ)in the 3-D T (*time*-space), where the bar under the symbol denote a vector with coordinates specified within the brackets ( ). In a 6-D ST, therefore, one can introduce the complete set of orthogonal *spherical polar* coordinates:

$$(r,\ \theta,\ \phi,\ t_r,\ t_\theta,\ t_\phi), \tag{2.9a}$$

where we have omitted for brevity the factor 'ic' associated with the time-coordinates. The above coordinates are of course interchangeable with the 6-D hyperspace *Cartesian* coordinates

$$(x_1,\ x_2,\ x_3,\ x_4,\ x_5,\ x_6), \tag{2.9b}$$

The following familiar interrelations provide inter-conversion between the two coordinate systems:

$$r = \left(x_1^2 + x_2^2 + x_3^2 \right)^{\frac{1}{2}} \tag{2.10a}$$

$$\theta = \cos^{-1}\left(\frac{x_3}{r} \right) \tag{2.10b}$$

$$\phi = \cos^{-1}\left[\frac{x_1}{\left(x_1^2 + x_2^2 \right)^{\frac{1}{2}}} \right] \tag{2.10c}$$

and, analogously,

$$t_r = \left(x_4^2 + x_5^2 + x_6^2 \right)^{\frac{1}{2}} \tag{2.10d}$$

$$t_\theta = \cos^{-1}\left(\frac{x_6}{t_r} \right) \tag{2.10e}$$

$$t_\phi = \cos^{-1}\left[\frac{x_4}{\left(x_4^2 + x_5^2 \right)^{\frac{1}{2}}} \right] \tag{2.10f}$$

As indicated before, for certain applications, it may be useful, for the sake of simplicity, to consider a 2-dimensional hyperspace (i.e., a "hyper-plane") constituted of the two "non-negative" coordinates, $r \geq 0$ and $t_r \geq 0$, the latter denoting the conventional time coordinate, associated with the "positive arrow" of time. It is readily evident that, for all practical purposes, Einstein's SRT equations for the space-time coordinate transformations, conveniently circumventing variations in two orthogonal spatial coordinates, y and z, are effectively confined to this (r, t_r) hyper-plane with the *time* coordinate t_r orthogonal to the radial *space* coordinate r, which essentially constitutes the direction of observation or of relative motion. Thus the 6-D ST hyperspace effectively reduces to the 4-D Einsteinian (SRT and GRT) spacetime, which is then further essentially reduced into the 2-D hyper-plane (r, t_r) for a spherically symmetric system. Symbolically,

$$(x_1, x_2, x_3, x_4, x_5, x_6) \rightarrow (x_1, x_2, x_3, x_4) = (x, y, x, ict) \rightarrow (r, t_r) \qquad (2.10g)$$

$$<....\, 6\text{-D ST} ...> <....\, 4\text{-D ST} ...> <..\, 2\text{-D ST} ..>$$

2.20. The Radial Separation in Terms of the Spherical Harmonics

The well-known fact that the strong interaction between two quarks increases with increasing separation between them naturally allows a simple modeling of the same in terms of a "spring"-like force or potential; and the related formulation of the strong interaction may be referred to as the 'spring theory.' The electromagnetic and gravitational interaction potentials vary in inverse proportion to the radial separation. Thus the radial separation between two points, specifying the locations of the two interacting systems (particles or cosmological bodies, for example) with respect to the observer is of prime interest in most cases.

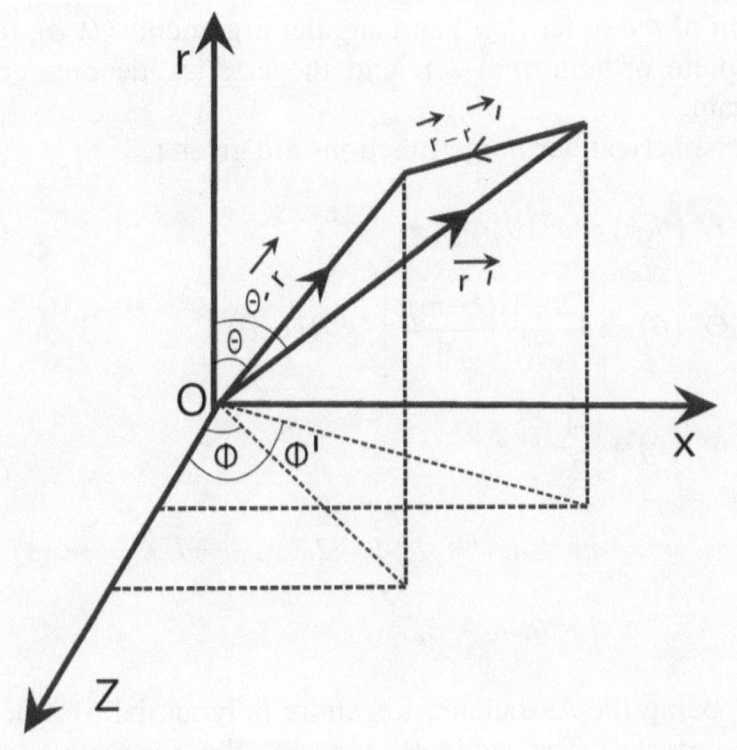

Figure 2.6   Spatial separation between two points in 3-D S

Here we indicate a general expression for the radial separation, r_s, between two points P (r, θ, ϕ) and P' (r', θ', ϕ') in a spherical polar coordinate system in the 3-D S in the form of a series expansion involving spherical harmonics[5]

$$r_s^{-1} = |\underline{r} - \underline{r}'|^{-1} = 4\pi \sum_{l=0}^{\infty} \sum_{m=-l}^{l} \frac{r_<^l}{r_>^{l+1}(2l+1)} Y_l^m(\theta,\phi) Y_l^{m*}(\theta',\phi'),$$

$$(2.11a)$$

where $\underline{r}$ and $\underline{r}'$ denote the vectors $\mathbf{OP}$ and $\mathbf{OP'}$, respectively, O being the origin of the 3-D S coordinate system (see Figure 2.6); r< and r> are the smaller and larger, respectively, of the lengths $|\mathbf{OP}|$ and $|\mathbf{OP'}|$; and $Y_l^m(\theta,\phi)$ is the spherical harmonic function of the order (l, m) and angular arguments (θ, ϕ), forming a complete orthonormal set; and the asterisk denotes complex conjugate.

The spherical harmonic functions are given as[5]

$$Y_l^m(\theta,\phi) = \Theta_l^m(\theta)\Phi_m(\phi) \tag{2.11b}$$

$$\Theta_l^m(\theta) = \left[\frac{(2l+1)(l-m)!}{2(l+m)!}\right]^{\frac{1}{2}} P_l^m(\cos\theta) \tag{2.11c}$$

$$\Phi_m(\phi) = \left\{\frac{1}{2\pi}\right\}^{\frac{1}{2}} e^{im\phi} \tag{2.11d}$$

$$m = -l, -l+1, -l+2, \ldots, +l \ ,$$

$$l = 0, 1, 2, 3, \ldots ,$$

$P_l^m(\zeta)$ being the Associated Legendre polynomial of order (l,m) and argument ζ. The angle θ_{12} between the pertinent vectors, **r** and **r'**, can also be written in terms of the series expansion:

$$P_l(\cos\theta_{12}) = \frac{4\pi}{(2l+1)}\sum_{m=-l}^{l} Y_l^{m*}(\theta,\phi) Y_l^m(\theta',\phi') \tag{2.11e}$$

where P_l is the Legendre polynomial of order l.

In analogy of Eqns. (2.11a) and (2.11e), we can also write, for the 3-D T space:

$$t_s^{-1} = |\underline{t}-\underline{t}'|^{-1} = 4\pi\sum_{l=0}^{l=\infty}\sum_{m=-l}^{m=l}\frac{t_<^l}{t_>^{l+1}(2l+1)}Y_l^m(t_\theta,t_\phi) Y_l^{m*}(t_\theta',t_\phi') \tag{2.12a}$$

$$P_l(\cos t_{\theta 12}) = \frac{4\pi}{(2l+1)}\sum_{m=-l}^{l} Y_l^m(t_\theta,t_\phi) Y_l^{m*}(t_\theta',t_\phi') , \tag{2.12b}$$

where the symbols have analogous meanings as in Eqns. (2.11a-e), but now they refer to corresponding variables in the time-space (3-D T).

In the 6-D ST hyperspace, the radial separation between the two points $P = (r, \theta, \phi, t_r, t_\theta, t_\phi)$ and $P' = \left(r', \theta', \phi', t'_r, t'_\theta, t'_\phi \right)$ can now be obtained using the mutual orthogonality of the spatial and time spaces. Applying the Pythagoras formula, this 6-dimensional "hyper-distance" between the points P and P' can be simply written as:

$$S' = \left[\left(r^2 + r'^2 - 2rr' \cos\theta_{12} \right) + \left(t_r^2 + t'^2_r - 2t_r t'_r \cos t_{\theta 12} \right) \right]^{\frac{1}{2}}$$

(2.12c)

The above results (Eqns. 2.11a-2.12c) are of fundamental and critical importance in interpreting observed experimental data in high-energy physics and cosmology. This is because, for two or more particles or bodies, the system behavior is governed by the relevant force-fields F(r) or, equivalently, by the potential functions V(r), [F(r) = $\partial V(r) / \partial r$] which might be dynamically changing in the 3-D S and the 3-D T spaces, i.e., in the 6-D ST hyperspace, due to their critical dependence on the instantaneous mutual separation, r, especially for a spherically symmetric system. However, the value of this separation is critically dependent on their relative location with respect to a particular observer, as shown by the referred set of equations (viz.,Eqns. 2.11a and 2.12a). This implies that, apart from the well known relativistic length-contraction and time-dilation effects (operative for a 'radially situated' observer), as predicted by SRT, there exists apparent variation of the pertinent space-time separation between the two interacting bodies, as interpreted by any observer, with explicit dependence on the bodies' coordinates with respect to the observer. A change of location of the observer with respect to the interacting system could thus produce an illusory effect indicating a different behavior pattern by the system. This happens because the angular coordinates change with a change in the location of the observer with respect to the two points interacting mutually, leading to a different value of the relative separation between them; and hence the values of the potential function and the force-field change accordingly, even though no such change should in principle occur in the system in absence of any *external* force. In particular, *two observers with distinct locations with respect to the system would interpret the data from the same event differently.* We term this effect "Observer-Location Dependent Effect (OLDE)." Clearly this effect can be quantitatively determined and compared with the value predicted from a theoretical analysis. In this connection it must be noted that the 'observer' could be a particle detector in a high-energy accelerator system, or a terrestrial or space-based telescope; and its location must be accounted for in interpreting the observed data. In case of a cosmological event, the variation in the location of the observer can obviously be considered relatively negligible due to the 'astronomical' distances involved; whereas it may play a critical role in the case of particle-accelerator related observations.

Equations 2.11(a)-2.12(c) also indicate the possibility of expressing the force-field F(r) or the potential function V(r) in terms of a series expansion in spherical harmonics, and thereby of finding possible solutions relevant in a unified theory of particles and interactions (UTOEPI). Clearly these solutions will explicitly depend on the coordinates of the objects under mutual interaction or of the interacting system (particles or cosmological bodies) in the frame of the observer. For the theory to be truly universal, this relativistic effect (OLDE) must be removed by using appropriate value of the potential based on universal constants.

2.21. Summary and Conclusions

A simple theoretical model of the Standard Model elementary particles and the four basic interactions, called the Unified Theory of Elementary Particles and Interactions (UTOEPI), is presented above, based on the concept of three-dimensional Time (3-D T), i.e., six-dimensional space-time (6-D ST) continuum—a paradigm shift from the conventionally accepted Einstein's 4-D ST continuum. Thus the proposed model demonstrates that Einstein's special relativity theory, being limited to only a single ('*imaginary*') time dimension (commonly denoted by ict) may present an incomplete picture of the physical reality in the domain of 'relativistic' velocity vector. This limitation, arising from the explicit assumption that the relative velocity between two frames of reference is aligned with a common axis of both the frames, is tantamount to treating the relative velocity as a scalar quantity. In effect, this simplification results in an inadvertent suppression of two other '*imaginary*' orthogonal time-dimensions. A more general treatment of the velocity as an arbitrary vector naturally leads to the concept of a three-dimensional time (3-D T) orthogonal to the three-dimensional space (3-D S) for a self-consistent theory. Consideration of this *complex* six-dimensional space-time (6-D ST) continuum, with the 3-D '*real*' spatial space and 3-D '*imaginary*' time-space, with the aid of the quaternion algebra, is seen to offer clues for solving many problems in elementary particle physics and cosmology; and lends itself toward the development of a simple model and unified theory of elementary particles and interactions (UTOEPI) presented in this Chapter.

One may also argue that there is no reason why a 3-D space (3-D S) should have only a single dimension of 'imaginary' time coordinate: The 3-D T could be said to be naturally expected for a complete symmetry between *spatial*-space and the *time*-space to form a proper space-time (6-D ST) continuum.

The primary implications of this 6-ST-based model can be best viewed in the framework of the quaternion algebra which is amply demonstrated here to physically represent rotation of a vector in a 4-D space constituted of *one real* and *three imaginary* dimensions. The set of successive rotations establishes the connection between such rotations with successive introduction of the complete set of interactions, and furnishes a natural explanation of the types and number of elementary particles in the Standard Model.

The set of successive rotations can be easily shown to correspond to introduction of the set of the four basic interactions. The model furnishes a simple, natural explanation of the types and number of elementary particles in the Standard Model. Furthermore, a simple relationship between the particle mass and its conserved interaction parameters is presented. As an important corollary, it can be said that the particle mass is a universal consequence of its interaction with the fields of interaction, each interaction adding, in a near-linear fashion (with one exception which leads to universal evolution of matter in the universe) to the mass of the particle. Last but not the least, this paper indicates scopes for obvious generalizations of the model presented here (UTEOPI) to include, if and as necessary, a self-consistent description (or prediction) of phenomena that could become observable in the future, at energy ranges of significantly higher orders of magnitude than presently available, such as the proposed Higgs field (the 'God-Particle') being sought by the LHC[4]. Finally, this model also leads to a simple explanation of the nuclear beta-decay and its important implications for the cosmological evolutionary transformations in the universe.

Thus here the concept of 6-D space-time (5-sphere) has been introduced as a simple applicable configuration for spherically symmetric system of elementary particles and basic Four Interaction fields (strong, weak, EM, and gravitational), in order to derive the (Standard Model) particle masses as a function of the field quantum parameters. The possibility of further enhancement and generalization of the model to include quaternion formulism, the Higgs field and supersymmetry have been briefly pointed out.

It is suggested that a similar sequence of phase transformations as simulated in this model might indeed have taken place in the cosmological evolution of the universe following the big bang. Last but not the least, this book indicates scopes for obvious generalizations of the UTEOPI model presented here to include, if and as necessary, a self-consistent description of phenomena that may become observable in the future, at energy ranges of significantly higher orders of magnitude than presently available. Such phenomena may including supersymmetry and the Higgs field (the 'God-Particle.') Additionally, further studies of the light cone, black holes, cosmic inflation, multiverse, cyclic universe, Anthropic principles, etc., may be carried out in the framework of the 6-D ST, contributing to a simple theory of the origin and evolution of the universe, free from unnecessary mathematical complexity, contrivances, and ambiguities.

References

[1]   O. Klein, Z. Phys., 37, 895 (1926)

[2]   G. A. Korn and T. Korn, Mathematical Handbook for Scientists and Engineers, (McGraw Hill, New York (1968), p. 871.

[3]   B. Carter, 'The Anthropic Principle and Its Implications for Biological Evolution,' Philosophical Trans. Of the Royal Society of London A 310 (1983), p. 347.

[4]   G. Kane, 'The Dawn of Physics Beyond the Standard Model,' Scientific American (June, 2003).

[5]   J. D. Jackson, Classical Electrodynamics, (John Wiley, New York (1962), p. 78

Further Reading

[1.a.] A. Einstein, *Einstein's 1912 Manuscript on the Special Theory of Relativity* (George Braziller, New York, 2003), p. 58, 80.

[b.]   W. Rindler, *Relativity: Special, General, and Cosmological* (Oxford University Press, Oxford, 2001).

[c.]   P.J.E. Peebles, *Principles of Physical Cosmology* (Princeton Univ. Press, Princeton, NJ, 1993), p. 267.

[2.a.] S. Hawking and R. Penrose, *The Nature of Space and Time* (Princeton Univ. Press, Princeton, NJ, 1996).

[b.]   P. Coveney and R. Highfield, *The Arrow of Time* (WH Allen, London, 1990), p. 157.

[3.a.]   O. Klein, Z. Phys. 37, 895 (1926).

[b.]   T. Kaluza, *Sitzungsberichte, Preussische Academie der Wissenschaften*, 966 (1921).

[4.]   P.J. Peebles, *Principles of Physical Cosmology* (Princeton Univ. Press, Princeton, NJ, 1993), p. 368.

[5.]   A. Einstein, H.A. Lorentz, H. Minkowski, and H. Weyl, *The Principle of Relativity* (1923), p. 104.

[6.]   J.D. Jackson, *Classical Electrodynamics* (John Wiley, New York, 1962), p. 78.

[7.]   G.A. Korn and T. Korn, *Mathematical Handbook for Scientists and Engineers* (McGraw Hill, New York, 1968), p. 871.

[8.]   Edited by G.W. Gibbons, E.P.S. Shellard, and S.J. Rankin, *The Future of Theoretical Physics and Cosmology* (Cambridge Univ. Press, Cambridge, 2003) p. 58.

[9.]   L. Hodderson, M. Brown, M. Riordan, and M. Dresden, *The Rise of the Standard Model; Articles on Particle Physics in the 1960s and 1970s* (Cambridge Univ. Press, Cambridge, UK, 1997).

[10.]   G. Kane, 'The Dawn of Physics Beyond the Standard Model,' Sci. Am. (June 2003).

[11.]   T. Van Flandern, *Dark Matter, Missing Planets and New Comets* (North Atlantic Books, Berkeley, CA, 1993), p. 47-52.

[12.a.]   P. Marcer, E. Mitchell, P. Rowlands, and W. Schempp, Zenergy: The 'Phaseonium' of Dark Energy that Fuels the Natural Structure of the Universe, International Journal of Computing Anticipatory Systems 16, 189-202 (2005).

[b.]   A. Challinor, 'Microwave Background Polarization in Cosmological Models,' astro-ph/9911481.

[c.]   W.L. Pritchard, H. Syderhoud and R. Nelson, *Satellite Communication System Engineering*, (Prentice Hall, NJ, 1993), p. 442, 454.

[d.]   M. Richharia, *Satellite Communication Systems*, (McGraw Hill, NY, 1999), p. 290.

[13.a.] A.H. Guth, 'Inflationary Universe: a possible solution to the horizon and flatness problem,' Phys. Rev. D 23, (1981), p.347-356.

[b.] A.H. Guth, *The Inflationary Universe* (Addison-Wesley, Reading, Mass., 1997); A.H. Guth, 'Eternal Inflation,' astro-ph/0101507.

[c.] A. Borde and A. Vilenkin, 'Eternal Inflation and the Initial Singularity, Phys. Rev. D 56, 717 (1997).

[d.] A. Borde, A.H. Guth, and A. Vilenkin, 'Inflation Is Not Past-Eternal,' gr-qc/011012.

[e.] A. Aquirre and S. Gratton, 'Steady State Eternal Inflation,' Phys. Rev. D 65, (2002), 083507.

[f.] J.M. Bardeen, P. Steinhardt, and M.S. Turner, 'Spontaneous Creation of Almost Scale-Free Density Perturbations in an Inflationary Universe,' Phys. Rev. D 28, 679 (1983).

[14.] L. Suskind and J. Lindesay, *An Introduction of Black Holes, Information and String Theory Revolution,* (World Scientific, NJ, 2005), p. 85.

[15.] H.R. Quinn and N. Yossi, *The Mystery of Missing Antimatter* (Princeton Univ. Press, Princeton, NJ, 2008).

[16.] S. Pakvasa, in *Neutrinos in the New Century, Particle Physics in the New Millenium,* proceedings of the 8th Adriatic Meeting (Springer, 2003), p. 50.

[17.] M. Veltman, *Facts and Mysteries in Elementary Particle Physics*, (World Scientific, NJ, 2003),

[18.] P. Rowlands, 'Zero to Infinity,' (World Scientific, NJ, 2008), p. 54.

[19.] G.A. Korn and T. Korn, *Mathematical Handbook for Scientists and Engineers* (McGraw Hill, New York, 1968), p. 385, 473.

[20.a.] J.B. Kuipers, *Quaternions and Rotation Squares* (Princeton Univ. Press, Princeton, NJ, 1999).

[b.] J.B. Kuipers, *Quaternions and Rotation Sequences: A Primer with Applications to Orbits, Aerospace, and*

Virtual Reality (Princeton Univ. Press, Princeton, NJ, 2002).

[c.] W. R. Hamilton, *Elements of Quaternions* (Longman, 1899).

[21.] G.D. Mahan, *Quantum Mechanics in a Nutshell* (Princeton Univ. Press, Princeton, NJ, 2009), p. 362.

[22.] J. Stillwell, *Yearning for the Impossible* (AK Peters, Ltd, Wellesley, Mass., 2006), p. 142.

[23.a.] G. Kane, *Supersymmetry* (Perseus Publishing, Cambridge, Mass., 2001), p. 68.

[b.] A. Seiden, *Particle Physics* (Addison Wesley, New York, 2005), p. 260.

[24.] Max Born, Atomic Physics, (Dover Publication, NY, 1969), p. 313.

[25.] V.J. Stenger, *The Comprehensible Cosmos*, (Prometheon Book, NY, 2006), p. 112.

[26.] V. Mukhanov, *Physical Foundations of Cosmology*, (Cambridge University Press, NY, 1969), p. 313.

[27.] B. Carter, 'The Anthropic Principle and Its Implications for Biological Evolution,' Philosophical Transactions of the Royal Society of London A 310, 347 (1983).

[28.] J. Gribbon and M. Rees, *Cosmic Coincidences: Dark Matter, Mankind, and Anthropic Cosmology* (Bantam Books, New York, 1989) chap. 10.

[29.a.] S.W. Hawking and G.F.R. Ellis, *The Large Scale Structure of Space-Time* (Cambridge Univ. Press, Cambridge, Mass., 1973).

[b.] T. Buchert and M. Carfora, 'Matter Seen at many scales and the Geometry of Averaging Relativistic Cosmology,' gr-qc/0101070.

[c.] R. Pensrose, 'Structure of Spacetime,' in Lectures in Matter and Physics, edited by C.M. DeWitt and J.A. Wheeler, (Benjamin, New York, 1968).

[d.] J.G. Muga, R. Sala Mayato, and I.L. Egusquiza, *Time in Quantum Mechanics*, Springer 2002; (For an extensive list of references, see J.J. Halliwell and J. Thorwart, (2002), gr-qc/0201070).

[e.]   G.W. Gibbons, E. Shellard and S. Rankin, Editors, *The Future of Theoretical Physics and Cosmology*, (Cambridge Univ. Press, UK, 2003).

CHAPTER 3

Matrix Representation of the Elementary Particle Mass-Interaction Relationship

"I have little patience with scientists who take a board of wood, look for its thinnest part, and drill a great number of holes where drilling is easy."

Albert Einstein (1879-1955)

3.1. Heuristic Parametric Modeling of Elementary Particles

It is now useful to seek a simple correlation among the basic parameters of the elementary particles, as summarized in Table 2.1. As mentioned before, these parameters include the mass (M), the strong charge (S), the weak charge (w), the electric (EM) charge (q), and the spin (s).

It may be noticed that spin of all the (Standard Model) elementary particles (all leptons or fermions, as per the particle statistics) has an identical common value (s = ½), and, therefore, for the present simplified modeling and analysis, the spin value could be considered a constant of the system of the elementary particles involved here. This would impact the assumption of the linear independence of the remaining parameters; and we will in fact represent the mass, M, in terms of a linear combination of the values of the other parameters, S, w, and q, for the simplified mathematical model here. The remaining four variables are obviously related to the four basic interactions, viz., gravitational (M), strong (S), weak (w), and electromagnetic (q), and in this section, we attempt to obtain a simple quantitative interrelation among there four variables (M, S, w, q).

Furthermore, based on the preceding discussions, the number of elementary particles considered for the present modeling in only nine (9), three particles belonging to each of the three Families, viz.

Family 1: $Q_u, Q_d,$ e⁻

Family 2: Q_c, Q_s, μ^-

Family 3: Q_t, Q_b, τ^- (3.1)

The three neutrinos may be considered separately in view of their unusual characteristics; viz., near-zero (or uncertain) mass, non-interaction with other particles and ordinary matter, and other strange properties. In particular, it is known that each of these neutrinos could be represented as a linear superposition of three more basic neutrino sub-states, associated with 'color' of the particles. One may be even tempted to postulate that the last characteristic might imply that the three types of neutrinos could in fact be three distinct states of one and the same particle. In other words, the three neutrino states could simply be manifestations of the neutrino-wave (i.e., the wave nature of the neutrino, according to the wave-particle duality principle) in the three orthogonal time coordinates, (t_1, t_2, t_3), respectively, although such an association is admittedly arbitrary at this stage. Extending this hypothesis to its extreme, one might even consider the possibility of the three Families of elementary particles as belonging to the three orthogonal time coordinates, such that only the Family 1 (F_1) particles correspond to the positive arrow of time coordinate of common day-to-day experience and observations; while the other two orthogonal time coordinates, to which Family 2 (F_2) and Family 3 (F_3) particles belong, keep them beyond ordinary observations, although they (F_2 and F_2 particles) might play a vital role in acting as the dark matter, as hypothesized earlier here. In any case, counting the neutrinos as one, the total number of elementary particles becomes equal to $N = 9 + 1 = 10$, as discussed above.

3.2.   The Mass Matrix and the Parameter Matrix

Thus, confining our consideration at the moment only to the nine elementary particles mentioned above (i.e., excluding neutrinos from our consideration for now, for simplicity), we notice that, while the masses of the particles are all different, the S, q and q parameter values of the three particles in each Family are the same. It is useful to introduce here a 3 × 3 'mass-matrix,' M, as well as a 3 × 3 'parameter-matrix,' T, as follows:

$$M = \begin{bmatrix} 0.0047 & 1.6 & 189 \\ 0.0074 & 0.16 & 5.2 \\ 0.00054 & 0.11 & 1.9 \end{bmatrix} \tag{3.2a}$$

$$T = \begin{bmatrix} 1 & \dfrac{1}{2} & \dfrac{2}{3} \\ 0 & \dfrac{1}{2} & -\dfrac{1}{3} \\ 0 & -\dfrac{1}{2} & -1 \end{bmatrix} \tag{3.2b}$$

The first column of the matrix represents the masses, in units of the proton mass, of the three elementary particles belonging to Family 1 (viz., Q_u, Q_d, e^-). Similarly, the second and third columns of the matrix represent the masses of the elementary particles of Family 2 and Family 3, respectively. Note that the numerical values of the elements of the mass-matrix would change depending on further accuracy in measurements and on the unit of representation for mass.

The first row of the matrix T represents the three parameter values (S, w, q) of the first particle-members of the three Families (F_1, F_2, F_3), i.e., of Q_u, Q_c, Q_t, respectively; similarly, the second row of the matrix T represent the (S, w, q) values of the second members of the three Families (F_1, F_2, F_3); and the third rows of the matrix T represent the (S, w, q) values of the third members of the three Families (F_1, F_2, F_3).

3.3.   The State Matrix

Finally, we introduce a 3×3 'state-matrix' of the elementary particles

$$\phi = \begin{bmatrix} \phi_{11} & \phi_{21} & \phi_{31} \\ \phi_{12} & \phi_{22} & \phi_{32} \\ \phi_{13} & \phi_{23} & \phi_{33} \end{bmatrix} \tag{3.2c}$$

and write the matrix relationship

$$M = T\phi \tag{3.3a}$$

i.e.,

$$M_{ij} = \sum_{k=1}^{3} T_{ik}\phi_{kj}, \qquad (i, j = 1, 2, 3) \tag{3.3b}$$

It then follows that the state matrix ϕ can be obtained from the matrix relation

$$\phi = T^{-1}M \tag{3.3c}$$

where T^{-1} is the 3×3 inverse matrix of the parametric-matrix;

$$TT^{-1} = T^{-1}T = I, \tag{3.3d}$$

I being the 3 × 3 identity matrix,

$$I = \begin{bmatrix} 1 & 0 & 0 \\ 0 & 1 & 0 \\ 0 & 0 & 1 \end{bmatrix}$$

(3.3e)

It is easy to evaluate the inverse matrix T^{-1} and the stat matrix ϕ using the relations postulated above. The results are

$$T^{-1} = \begin{bmatrix} \dfrac{4}{3} & -\dfrac{1}{3} & 1 \\ -2 & 2 & -2 \\ 1 & -1 & 0 \end{bmatrix}$$

(3.4a)

$$\phi = \begin{bmatrix} 0.000029 & 0.767778 & 84.166667 \\ -0.00432 & -3.1 & 371.4 \\ -0.0027 & 1.44 & 183.8 \end{bmatrix}$$

(3.4b)

3.4.   Mass-Interaction Relation

The matrix Equation (3.3a), together with the definitions of the mass-matrix M (Eqn. 3.2a), the parameter matrix T (Eqn. 3.2b) and the state matrix ϕ (Eqn. 3.4b), provides the desired interrelations between the fundamental parameters (S, w, q, and M) of the main elementary particles of the Standard Model, as related to the Four basic interactions (strong, weak, electromagnetic and gravitational). In this simplistic, heuristic model of elementary particles and the basic interactions, the physical significance of the state matrix (ϕ), possibly related to the Higgs field, as well as suitable incorporation of the neutrino, have not been explicitly addressed. Also, clearly, in this model, mass has been assumed to be linearly contributed by the strong, weak, EM interaction fields, which may be an oversimplification. These and other related aspects of the model, including the method to generalize it to higher dimensions (say, up *to 11 dimensions, consisting of 6 dimensions of the 6-D ST together with 5 dimensions for the independently treated parameters, S, w, q, M and s*) could be investigated further. Such an approach could also be possibly applied for modeling supersymmetry, if desired. Additional variations in the model could include a nonlinear (say, quadratic) dependence of mass on other parameters and use of the quaternion formulism, as indicated above.

CHAPTER 4

Relativistically Covariant Representations of Particle Distribution and Interaction Fields

"I am a great believer in the simplicity of things....I am inclined to hang on to board simple ideas like grim death until evidence is too strong for my tenacity."

Ernest Rutherford (1871-1937)

4.1. Covariat and 6-Covariant Equations

In Chapter 4, we consider the equations satisfied by the particle distributions and the interaction fields that can best describe the primordial matter regarded as fluid. Starting with the Boltzmann equation, first a four-covariant equation is derived in the context of the Einsteinian four-dimensional spacetime (4-D ST) continuum of the special relativity theory (SRT). It is then generalize for the six-dimensional spacetime (6-D ST) of the extended relativity theory (ERT) propounded here. Again, simplicity of mathematical formulation and presentation is kept in mind as a prime factor.

4.2.   Unified Equations for Particles (Matter) and Interactions (Fields)

It has been concluded by many investigators that the ultra-high energy 'soup' of the quarks and gluons immediately following the big bang (on the time-scale of the Planck time) behaved much like a fluid under a unified interaction. Subsequently, of course, cosmic eras of bifurcations of the strong, weak, electromagnetic (EM) and gravitational interactions followed with evolutionary cooling of this soup at various characteristic temperatures and cosmological expansion phases accompanied with onset of nucleo-synthesis, nuclear decay, formation of electron-ion pairs (ionic plasma and neutral atoms, stars, galaxies, clusters and super-clusters, etc.) at different stages of the evolution of the universe. Thus, it is useful to explore the possibility of describing the behavior of matter in fluid form in relativistic high temperature and velocity ranges in the form of the quark-gluon soup, relativistic plasma, and neutral particle stages in order to understand the overall cosmological evolution based on the interactive field and particle distributions.

A most convenient and suitable type of mathematical equation to describe fluids or a statistical populations of material particles is the Boltzmann-Maxwell equation, reducing to the Vlasov equation when particle collision interactions become negligible. The Boltzmann and Vlasov equations lend themselves to treatment of electromagnetic forces (electric and magnetic fields or, equivalently, vector and scalar potentials, in the presence of electrically charged particles.) Generalizations of the Boltzmann or Vlasov equation to treat quark-gluon, relativistic plasma, radioactive and neutral particle and field distributions should also be possible. However, for a general applicability and universality of the Boltzmann and Vlasov equations, the non-relativistic form of the conveniently employed equations (to describe ordinary fluids, plasmas, and particle ensembles) must first be cast into a relativistic manifestly covariance forms, in the four-dimensional spacetime (4-D ST) and in the six-dimensional spacetime (6-D

ST), so that the laws of physics retain the same form in reference frames moving with respect to one another.

In the following Sections, first a 4-covariant form (in the 4-D ST) of the Maxwell-Boltzmann (or Vlasov) equation is derived as a first-order approximation for application to high-temperature (relativistic) plasmas, treated as a system subject to the special relativity theory (SRT) and EM interaction. Then this equation is extended to assume a manifestly covariant form for the case of a six-dimensional space-time (6-D ST) involving the introduction of a three-dimensional orthogonal time-space (3-D T), in addition to the usual three-dimensional space (3-D S); that is a 6-covariance, for possible application for all forms of matter and fields, including the quark-gluon soup, elementary particle including neutrino population, high-energy relativistic plasma, and cosmological matter, treated as a system subject to the extended relativity theory (ERT) and various relevant interactions.

The above type of a formulation could thus also be useable for studying likely evolutionary patterns and forms of the primordial matter immediately following the big bang. Passages or phase-changes of this matter through various cosmic structures could thus be studied and examined in a unified manner, with all particles and bodies together with their interaction fields.

4.3.   Manifestly 4-Covariant Formulation of the Boltzmann-Vlasov Equations

In terms of the conventional 4-D field vectors $A_\mu = \left(\vec{A}, i\varphi\right)$, the Boltzmann equation for distribution $f\left(\vec{x}, \vec{V}, t\right)$ of an ensemble of particles of mass m and change e is shown below:

$$\frac{\partial f}{\partial t} + \vec{V}\cdot\frac{\partial f}{\partial \vec{x}} + \frac{e}{m}\cdot\left\{-\vec{\nabla}\varphi - \frac{1}{c}\cdot\frac{\partial \vec{A}}{\partial t} + \frac{\vec{V}\times\vec{\nabla}\times\vec{A}}{c}\right\}\cdot\frac{\partial f}{\partial \vec{V}} = \left(\frac{\partial f}{\partial t}\right)_c$$

(4.1a)

Here, $\left(\dfrac{\partial f}{\partial t}\right)_c$ is a proper collision integral.

The relativistic rest-mass m_0 is given by

$$m = \gamma_0 m_0$$

$$\gamma_0 = \frac{1}{\sqrt{1 - \dfrac{V^2}{c^2}}}$$

where c is the velocity of light, and the improper four-velocity vector is

$$V_\mu = \left(\vec{V}, ic\right).$$

Also, note that

$$\gamma_0 = \frac{c}{i\sqrt{V^2 + \left(ic\right)^2}} = -V_4 \left(V_\mu \cdot V_\mu\right)^{-\frac{1}{2}}$$

where the summation convention (implying summation over repeated indices) is employed. We can now rewrite Eqn. (4.1a) in the form[1],

$$V_\mu \partial_\mu f + \frac{e}{\gamma_0 m_0} \left[i\partial_k A_4 - i\partial_4 A_k + \frac{1}{c}\left(V_l \partial_k A_l - V_l \partial_l A_k\right)\right] \frac{\partial f}{\partial V_k} = \left(\frac{\partial f}{\partial t}\right)_c$$

i.e.,

$$V_\mu \partial_\mu f + \frac{e}{\gamma_0 m_0 c}\left\{\left(V_4 \partial_k A_4 + V_l \partial_l A_l\right) - \left(V_4 \partial_k A_4 + V_l \partial_k A_l\right)\right\}\frac{\partial f}{\partial V_k} = \left(\frac{\partial f}{\partial t}\right)_c$$

or[2]

---

[1] Since,
$$\left(\vec{V} \times \vec{\nabla} \times \vec{A}\right)_{ic} = \varepsilon_{klm} V_l \left(\nabla \times \vec{A}\right)_m = \varepsilon_{klm} V_l \varepsilon_{mnp} \partial_n A_p = \varepsilon_{klm}\varepsilon_{mnp} V_l \partial_n A_p = \left(\delta_{ln}\delta_{lp} - \delta_{kp}\delta_{ln}\right)V_l \partial_n A_p = V_l \partial_k A_l - V_l \partial_l A_k$$

[2] Since, $\dfrac{\partial f}{\partial V_4} = 0, V_4 = ic$ being an invariant quantity

$$V_\mu \partial_\mu f + \frac{e}{\gamma_0 m_0 c}\left(V_\mu \partial_\nu A_\mu + V_\mu \partial_\mu A_\nu\right)\frac{\partial f}{\partial V_\nu} = \left(\frac{\partial f}{\partial t}\right)_c \tag{4.1b}$$

We must now convert V_μ into a proper 4-vector (e.g. $p_\mu = mV_\mu$). Noting first that[3]

$$\frac{\partial f}{\partial V\nu} = \frac{\partial f}{\partial(\gamma_0 V_\lambda)}\frac{\partial(\gamma_0 V_\lambda)}{\partial V_\nu} = \left(\gamma_0 \frac{\partial V_\lambda}{\partial V_\nu}\right)\frac{\partial f}{\partial(\gamma_0 V_\lambda)} = \gamma_0\left\{\frac{\partial f}{\partial(\gamma_0 V_\lambda)} + \frac{V_\lambda \gamma_0^2 V_\nu}{c^2}\frac{\partial f}{\partial(\gamma_0 V_\lambda)}\right\}$$

We have, by substituting the above in equation (4.1b) and rearranging[4],

$$(\gamma_0 V_\mu)\partial_\mu f + \frac{e}{m_0 c}\left\{(\gamma_0 V_\mu)\partial_\nu A_\mu - (\gamma_0 V_\mu)\partial_\mu A_\nu\right\}\left\{\frac{\partial f}{\partial(\gamma_0 V_\nu)} + \frac{(\gamma_0 V_\lambda)(\gamma_0 V_\nu)}{c^2}\frac{\partial f}{\partial(\gamma_0 V_\nu)}\right\} = \gamma_0\left(\frac{\partial f}{\partial t}\right)_c$$

i.e.

$$p_\mu \partial_\mu f + \frac{e}{c}p_\mu F_{\nu\mu}\left\{\frac{\partial f}{\partial p_\nu} + \left(\frac{p_\lambda p_\nu}{m_0^2 c^2}\right)\frac{\partial f}{\partial p_\lambda}\right\} = m_0\left(\frac{\partial f}{\partial \tau}\right)c$$

where we have used the familiar 4-vector and tensors:

$$p_\mu = \gamma_0 m_0 V_\mu = \left(\vec{p}, \frac{i\varepsilon}{c}\right);$$

$$\vec{p} = \gamma_0 m_0 \vec{V}$$

$$\varepsilon = mc^2 = \gamma_0 m_0 c^2 = \sqrt{p^2 c^2 + m_0^2 c^4}$$

with,

---

[3] $\gamma_0 = -ic\left(V_\mu V_\mu\right)^{-\frac{1}{2}}$

Hence, $\dfrac{\partial \gamma_0}{\partial V_\nu} = \dfrac{ic}{2}\left(V_\mu V_\mu\right)^{-\frac{3}{2}}\cdot 2V_\mu \delta_{\mu\nu} = \dfrac{\gamma_0^3}{c^2}V_\nu$, and also, $\dfrac{\partial \gamma_\lambda}{\partial V_\nu} = \delta_{\lambda\nu}$

[4]   If any quantum operator formulism be invoked, then here we have assumed that V_μ or γ_0 commutes with $\partial_\mu A_\nu$.

$$p = \left|\vec{p}\right|, \; p_\mu p_\mu = -m_0^2 c^2 \, ,$$

and,

$$F_{\nu\mu} = \partial_\nu A_\mu - \partial_\mu A_\nu \tag{4.1c}$$

together with proper time τ, where

$$d\tau = \sqrt{1 - \frac{v^2}{c^2}} dt$$

i.e.,

$$\gamma_0 \frac{\partial}{\partial t} = \frac{\partial}{\partial \tau}$$

By virtue of the antisymmetric nature of the electromagnetic tensor $F_{\mu\nu}$, the term containing $\frac{\partial f}{\partial p_\lambda}$ drops out, so that we have the desired manifestly covariant form of the Boltzmann equation:

$$p_\mu \partial_\mu f + \frac{e}{c} p_\mu F_{\nu\mu} \frac{\partial f}{\partial p_\nu} = m_0 \left(\frac{\partial f}{\partial \tau} \right)_c \tag{4.2a}$$

and, for the collisionless case, the covariant Vlasov equation:

$$p_\mu \partial_\mu f + \frac{e}{c} p_\mu F_{\nu\mu} \frac{\partial f}{\partial p_\nu} = 0 \tag{4.2b}$$

Now, in obtaining Eqns. (4.2a,b), we have, in effect, added extra terms to the original (non-relativistic) Vlasov equation, which is the terms (added under the "assumption" that $\frac{\partial f}{\partial V_4} = 0$).

$$\frac{e}{c} p_\mu F_{4\mu} \frac{\partial f}{\partial p_4} = ie\left(\vec{p}, \vec{E}\right)\frac{\partial f}{\partial \varepsilon} = i\gamma_0 e m_0 \left(\vec{V}, \vec{E}\right)\frac{\partial f}{\partial \varepsilon} = i\gamma_0 \frac{e}{c^2}\left(\vec{V}, \vec{E}\right)\frac{\partial f}{\partial \gamma_0}$$

$$\tag{4.3}$$

where we have used the results: $F_{4\mu} = (-E_z, B_y, -B_x, 0)$; since in the non-relativistic limit,

$$\gamma_0 = \frac{1}{\sqrt{1-\dfrac{V^2}{c^2}}}$$

becomes a constant, the above term drops off and Eqns. (4.2 a,b) reduce to the old form. This new term, as shown in Eqn. (4.3) is proportional to the power spent by the electric field in particle motion. As power is the fourth component of the covariant 4-force, the above addition to the Boltzmann-Vlasov equation for 4-covariance is just expected.

For reference purposes, the well-known (4 × 4) matrix element representation of the tensor, of order 2 in terms of the three dimensional electromagnetic (EM) field-components in the 4-D ST (in Cartesian coordinates, x, y, z, ict) is given below:

$$F_{\mu\nu} = \begin{bmatrix} 0 & Ex & Ey & Ez \\ -Ex & 0 & Bz & -By \\ -Ey & -Bz & 0 & Bx \\ -Ez & By & -Bx & 0 \end{bmatrix} \tag{4.4}$$

4.4. Manifestly 6-Covariance: Generalization of the Equation to the 6-D Space-time

A generalization of the 4-D covariant Boltzmann-Vlasov Eqn. (4.2 a,b) to the case of the 6-D ST is obvious: while this generalized Equation can still formally be written in the same from as Eqn. (4.2 a,b), now the tensor indices, (μ,ν), are not limited to the values (1, 2, 3, 4), but must take the values $(\mu,\nu) = $ (1, 2, 3, 4, 5, 6). The first three indices $(\mu,\nu = 1, 2, 3)$ refer to the tree-dimensional space (3-D S), and the remaining three indices $(\mu,\nu = 4, 5, 6)$ refer to the three-dimensional time (3-D T) space. The 4-D ST tensor and the matrix $F_{\mu\nu}$, Eqn. (4.4), is now to be extended to a new, (6 × 6), matrix, $G_{\mu\nu}$, which can be formally represented, in terms of the more general 6-D ST notation $(x_1, x_2, x_3, x_4, x_5, x_6)$, as:

$$
G_{\mu\nu} = \left[
\begin{array}{cccc|cc}
0 & E_1 & E_2 & E_3 & G_{15} & G_{16} \\
-E_1 & 0 & B_3 & B_2 & G_{25} & G_{26} \\
-E_2 & -B_3 & 0 & B_1 & G_{35} & G_{36} \\
-E_3 & B_2 & -B_1 & 0 & G_{45} & G_{46} \\
\hline
G_{51} & G_{52} & G_{53} & G_{54} & 0 & G_{56} \\
G_{61} & G_{62} & G_{63} & G_{64} & G_{65} & 0
\end{array}
\right] \tag{4.5}
$$

Clearly, $G_{\mu\nu}$, with $\mu,\nu = 1, 2, 3, 4, 5, 6$ is composed of four component matrices: a (4 × 4; upper left) matrix, identical to $F_{\mu\nu}$ of Eqn. (4.4); a (4 × 2; upper right) matrix, say, $S_{\mu\nu}$; a (2 × 4; lower left) matrix, say, $W_{\mu\nu}$; and a (2 × 2; lower right) matrix, say, $g_{\mu\nu}$, composed of two non-zero elements G_{56} and G_{65}.

Just as $F_{\mu\nu}$ contains components of the electromagnetic (EM) interaction (and the EM carrier, the photon), the remaining three matrices could be expected to bear simple relationships with the other three basic interactions—strong, weak and gravitational—and the corresponding carrier fields (gluon, weak bosons, and gravitons). In such a case, the above model or representation could constitute a simple unified theory of all the four basic interactions. The unified interactions can thus be said to be represented by the 6 × 6 matrix, $G_{\mu\nu}$, which can be simply partitioned as described above to represent the four basic interactions. By virtue of the defining equation, (analogous to Eqn. (4.1c), but with $\mu,\nu = 1, 2, 3, 4, 5, 6$), the fields for the strong, weak and gravitational interactions could then be directly correlated with the partial differentials with respect to the extra time-dimensions, t_2 and t_3, respectively, introduced in this book, leading to a 6-D ST continuum-based model of a unified theory of elementary particles and interactions (UTOEPI). Thus, through the above generalization, the Boltzmann-Vlasov equation has been extended to treat ensembles of quark-gluon plasma and weak-interaction (decay) of neutral particles as well. The above characterization of the sub-matrices for the strong, weak, and gravitational interaction is somewhat tentative, however.

Potentially the most central and critical role of the additional time dimensions (t_2, t_3) evidently lies in the fact that, among other factors, the associated equations also satisfy a simple (and obviously required) symmetry in the space and time coordinates, and are amendable to an arbitrary value and orientation of the relative velocity vector with respect to the direction of the observation (see Eqn. 1.1e). This facet is crucial for a realistic formulation and analysis of high-energy (Planck-scale) elementary particle physics and cosmology where it is not possible to have the direction of observation coincide with the direction of the relative motion of particles and systems involved, as may be the case in laboratory experiments or certain simple astrophysical observations.

4.4. Summary

As a generalization of the conventional Boltzmann-Vlasov equations describing the dynamics of the particle and field distributions, a manifestly 4-covariant and a manifestly 6-covariant (in the 6-dimensional space-time continuum propounded in this book) forms of the equations are derived. It is suggested that the equations could simply relate to the matrix representation of the pertinent interactions (strong, weak, electromagnetic, and gravitational) in a unified manner. Its relevance and applications for the unified theory of elementary particles and interactions (UTEOPI) as well as for other pertinent systems are briefly pointed out.

CHAPTER 5-A

An Algebraic Unified Representation of the Interaction Potentials

5.1. Potential Function Representation

The basic interactions can of course be represented in terms of suitably defined potential functions of the coordinate variables. Here we review the well-known interaction potentials in the 6-D ST, and use the same to construct quaternions which, in turn, could be used to derive expressions for the elementary particle mass in terms of knows interaction constants, or vice-versa.

In general, the force experienced by a particle or body could be simply written as

$$F\left(x\right) = \frac{\partial V\left(x_1, x_2, ..., x_6\right)}{\partial}, i = 1, 2, ..., 6 \tag{5.1}$$

where μ is the index representing the type of interaction, x_i (i = 1,2, . . . ,6) is the ith coordinate, V_μ is the potential function and $F_\mu(x_i)$ is the force component in the direction of x_i. We now review the potential functions for the strong ($\mu = 1$), weak ($\mu = 2$), EM ($\mu = 3$) and gravitational ($\mu = 4$) interaction and indicate uniformity involved therein.

For spherically symmetric case applicable for basic interactions, only the radial spatial and time coordinates (r,t_r) are involved, so that the potential functions are of the forms $V(r)$ and $V(t)$ only, where we have used the notation $t = t_r$ for simplicity.

5.2. Strong Interaction

Using the Yukawa-type of model, and introducing the strong interaction charge (S), we can simply write

$$V_1(r) = -SV_{1,0}e^{-\frac{r}{r_n}} \qquad\qquad (5.2a)$$

or

$$V_1(r) = V_1'Se^{-c_1 r^{b_1}} \qquad\qquad (5.2b)$$

where $V_{1,0}$ is the depth of the potential well, and r_n is the radius of the nucleon or the radius of the nucleus of a radioactive element involved; so that the quarks or the nucleons are bound with a maximum strength $V_{1,0}$ (binding energy) within the radius r_n under the strong interaction, diminishing exponentially for $r > r_n$, and

$$V_1 = -V_{1,0} \qquad\qquad (5.2c)$$

$$b = 1 \qquad\qquad (5.2d)$$

$$c_1 = \frac{1}{r_n} \qquad\qquad (5.2e)$$

Expanding the exponential in Eqn.5.2(a), and taking partial differential with respect to the radial separation r, we have the (radial) strong force given as (S = 1)

$$F_1(r) = \frac{\partial V_1(r)}{\partial r}$$

$$= -V_{1,0} \frac{\partial}{\partial r}\left(1 - \frac{r}{r_n} + \frac{r^2}{2r_n^2} - \frac{r^3}{6r_n^3} + \ldots\right)$$

$$= \frac{V_{1,0}}{r_n}\left(1 - \frac{r}{r_n} + \frac{r^2}{2r_n^2} - \ldots\right)$$

$$\simeq F_{1,0} - kr + k'r^2 - \ldots \qquad (5.3a)$$

where

$$F_{1,0} = \frac{V_{1,0}}{r_n} \qquad (5.3b)$$

$$k' = \frac{V_{1,0}}{2r_n^3} \qquad (5.3c)$$

$$k = \frac{V_{1,0}}{r_n^2} \qquad (5.3d)$$

Thus, apart from the constant term $(F_{1,0})$, the strong (nuclear) force primarily exhibits an elastic spring-force like characteristics—being an attractive force increasing with the separation (distance $r < r_n$) between the interacting quarks (or nucleons), the effective "strong interaction elastic constant" (SIEC) varying inversely with the square of the nucleon radius and directly with the binding energy $(V_{1,0})$. For example, the SIEC-value is much larger for quarks (r_n= the neutron radius) than for heavy radioactive nuclei.

The second order term ($\propto r^2$) repulsive in nature and proportional to the coefficient k', can be thought of as responsible for, or contributing to, the β-decay or nuclear decay for large enough separation ($r < r_n$), thereby overcoming the strong interaction (attractive or binding force.) This type of effect may therefore lead to the interpretation of the strong and weak interactions as two sides of the same coin. The weak interaction-induced neutron—or radioactive nuclear—decay is thus analogous to "snapping of the spring" under excessively stretched length (inter-quark or inter-nucleon separation distance.) In view of the above observation, the strong and weak interaction phenomena could be regarded as complementary and being described by the unified 'spring theory' (in contrast to the popular 'string theory.') This view is re-enforced by noting that the weak interaction-induced decay formula involves exponential function similar to the exp(-r/r_n), but with the spatial radial separation being replaced by the radial "separation" $t_r = t$ (say) in the time-domain, as indicated below.

5.2. Weak Interaction

The familiar neutron or nuclear decay formula

$$N(t) = N_0 e^{-\frac{t}{\tau}} \tag{5.4a}$$

where $N(t)$ is the number of nucleons at time t, N_0 being the initial number and τ the half-life involved, could be rewritten in terms of the fourth coordinate in the 4-D (or 6-D) space-time

$$x_4 = ict_r = ict \tag{5.4b}$$

$$x_4' = ic\tau \tag{5.4c}$$

(with $i = \sqrt{(-1)}$)

Thus, introducing a weak interaction potential function $V_2(x_4)$ assumed to be proportional to the instantaneous number (N) of neutrons or nucleons, we can write

$$V_2(x_4) = \omega V_{2,0} e^{-\frac{x_4}{x_4}}$$
(5.4d)

where $V_{2,0}$ is the initial value of the potential corresponding to the initial number N_0, and we have explicitly introduced the weak interaction charge, ω, on the right of Eqn. 5.4(d). The effective "weak force" can be obtained by differentiating $V_2(x_4)$ with respect to the variable x_4 (i.e., with respect to time t).

In analogy with the strong interaction potential function (Eqn. 5.2b), the weak interaction potential can be simply rewritten as

$$V_2(x_4) = \omega V_2' e^{-c_2 x_4^{b_2}}$$
(5.4e)

where

$$V_2' = V_{2,0}$$
(5.4f)

$$b_2 = 1$$
(5.4g)

$$c_2 = \frac{1}{x_4'} = \frac{1}{ic\tau} = \frac{-i}{r_2}$$
(5.4h)

and $r_2 = c\tau$ is the distance traversed by light during the time interval τ, the decay half-life.

5.3. Electromagnetic (EM) Interaction

The familiar EM potential is

$$V3(r) = -\frac{\alpha q_0 q}{r}$$
(5.5a)

yielding the EM force varying inversely as the square of distance r

$$F_3(r) = \frac{\partial V_3(r)}{\partial r} = \frac{\alpha q_0 q}{r^2}$$
(5.5b)

where q_0 is the electric charge, in units of the electron charge ($|e|$), of the particle or body producing the EM force-field, q is the electric (i.e., EM) charge of the particle or body being examined, also in units of $|e|$, and α is constant depending on units.

The minus sign chosen for $V_3(r)$ (Eqn. 2.5a) ensures that the EM force is repulsive if q_0 and q are of the same sign (both positive or both negative), and attractive if q_0 and q bear opposite signs.

The potential $V_3(r)$ can be rewritten in a form analogous to that of the strong and weak interaction, as follows

$$V_3\left(r\right)=V_3'qe^{-c_3 r^{b_3}} \tag{5.5d}$$

where

$$V_3'=-\frac{\alpha q_0}{r_N} \tag{5.5e}$$

$$b_3 = -1 \tag{5.5f}$$

$$c_3 = -r_N \tag{5.5g}$$

where it is assumed that the EM force primarily and effectively operates in the range $r > r_N$, where r_N is the radius of the nucleus.

Using Eqn. 5.5(f) and 5.5(g), the exponential term arbitrarily introduced on the right Eqn. 2.5(d) is seen to be

$$e^{+\frac{r_N}{r}}=1+\frac{r_N}{r}+\frac{1}{2!}\left(\frac{r_N}{r}\right)^2+\frac{1}{3!}\left(\frac{r_N}{r}\right)^3+... \tag{5.6a}$$

so that, in the range $r > r_N$, the quadratic and higher order terms in the above series expansion progressively become relatively negligible; and retaining the first order term normally suffices. Thus, Eqn. 5.5(d) can be approximately written as

$$V_3(r) \simeq -\frac{\alpha q_0 q}{r_N}\left(1+\frac{r_N}{r}\right)$$

$$(5.6b)$$

Differentiating both sides of Eqn. 5.6(b) with respect to r we reproduce the EM force equation (Eqn. 5.5b), validating the postulate of Eqn. 5.5(d-g) for the modified EM potential function. It may now also be postulated that the potential function (5.5d) for $r < r_N$ represents large negative value, although the corresponding force value is relatively negligible compared to the strong (and effective weak force) in this range. Similarly, inclusion of the higher order terms in Eqn. 5.6(a) should, in principle, correspond to higher order "corrections" to the EM force as given by the Coulomb law, though for all practical purposes, such higher order corrections are negligible.

5.4.   Gravitational Interaction

Starting with Newton's law of gravitation for a particle or body of mass m,

$$V_4(r) = -\frac{Gm_0 m}{r^2} = \frac{\partial V_4(r)}{\partial r}$$

$$(5.7a)$$

and the corresponding potential function

$$V_4(r) = \frac{Gm_0 m}{r}$$

$$(5.7b)$$

where m_0 is the mass of the body creating the gravitational force-field and G is the universal gravitational Constant. Let us introduce the modified gravitational potential in a manner similar to the preceding cases:

$$V_4(r) = V_4' m e^{-c_4 r^{b_4}}$$

$$(5.7c)$$

where

$$V_4' = \frac{Gm_0}{r_a} \tag{5.7d}$$

$$b_4 = -1 \tag{5.7e}$$

$$c_4 = -r_a \tag{5.7f}$$

where r_a is the atomic/molecular radius. The exponential function on the right of Eqn. 5.7(c) is

$$e^{\frac{r_a}{r}} = 1 + \frac{r_a}{r} + \frac{1}{2!}\left(\frac{r_a}{r}\right)^2 + \frac{1}{3!}\left(\frac{r_a}{r}\right)^3 + \dots \tag{5.8a}$$

so that, in the range $r > r_a$, the quadratic and higher order terms in the above series expansion are relatively negligible, and retaining the first order term suffices. Thus, Eqn. 5.7(c) can be approximately written as

$$V_4(r) \simeq \frac{Gm_0 m}{r_a}\left(1 + \frac{r_a}{r}\right) \tag{5.8a}$$

Differentiating both sides of Eqn. 5.8(a), we can reproduce Eqn. 5.7(a), validating the choice of the modified gravitational potential (Eqn. 5.7c). The value of this potential for $r < r_N$ corresponds to negligibly small force compared to the relatively much larger strong-, weak- and EM-forces in this region, whereas inclusion of the quadratic and higher terms of Eqn. 5.8(a) should, in principle, represent "corrections" to the basic (Newton's) law of gravitation (Eqn. 5.7a).

5.5. Unification of Interactions: Summary of Results

From Eqns. 5.2(b), 5.4(e), 5.5(d) and 5.7(c), we can write the general expression of the potential function for the μth interaction as

$$V_\mu = V'_\mu p_\mu e^{-c_\mu (x_\mu)^{b_\mu}}, \mu = 1,2,3,4 \tag{5.9}$$

where V'_μ represents the "interaction strength-factor," p_μ the "interaction charge," and c_μ and b_μ are additional model parameters for the μth interaction (recall: μ = 1,2,3,4 for the strong, weak, EM and gravitational interaction, respectively.) Table 5.1 summarizes these factors, charges and parameters for convenience of reference.

Table 5.1—Summary of Results for Interaction Potential Function V_μ (Eqn. 5.9)

Value of μ	V_μ	Type of Interaction	Coordinate Variable	Interaction Strength Factor (V'_μ)	Interaction Charge (p_μ)	Parameters	
						b_μ	c_μ
1	V_1	strong	r	$-V_{1,0}$	S	1	$\dfrac{1}{r_n}$
2	V_2	weak	x_4	$V_{2,0}$	ω	1	$\dfrac{1}{ic\tau}$
3	V_3	Electromagnetic (EM)	r	$-\dfrac{\alpha q_0}{r_N}$	q	-1	$-r_N$
4	V_4	Gravitational	r	$\dfrac{Gm_0}{r_a}$	m	-1	$-r_a$

5.6. The Composite Interaction Potential

We can now define the composite or the grand total potential function (GTPF), V_T, simply as the sum total of the potential functions for the four types of interactions

$$V_T = V_1 + V_2 + V_3 + V_4 \tag{5.10a}$$

$$\sum \tag{5.10b}$$

i.e.,
$$V_T = \sum_{\mu=1}^{4} V'_\mu p_\mu e^{-c_\mu \left(x_\mu^{b_\mu}\right)} \tag{5.10c}$$

[where, as used in Eqn. 5.10(c) for brevity, we must use $x_1 = x_2 = x_3 = r$, the radial separation between interacting particles or bodies, and $x_4 = ict$.] All the symbols involved are summarized in Table 5.1. We can now examine the consequences of two hypotheses: (a) $V_1 = V_2 = V_3 = V_4$ (Potential energy conservation principle), and (b) $V_T = 0$ (minimum potential principle.) The composite potential could be appropriately used in the radial portion of the Schrodinger equation and solved numerically, in case an analytical solution become intractable.

5.7. Quaternion for the Composite Interaction

It may be useful to define now the grand quaternion (GQ) for the composite interaction, Q_T, as follows

$$Q_T = V_4 + iV_1 + jV_2 + kV_3 \tag{5.11a}$$

where V_μ ($\mu = 1,2,3,4$) are real potential function defined by Eqn. (5.9), together with Table 5.1; and (i, j, k) are quaternion generators (see Section 2.8.) The interpretation of Q_T as composed of four components, each corresponding to a particular interaction-induced phase-change (rotation by $\frac{\pi}{2}$) in the 6-D ST phase-space can again be invoked as before. The magnitude of the quaternion Q_T is given by

$$|Q_T| = \left[(V_4)^2 + (V_1)^2 + (V_2)^2 + (V_3)^2 \right]^{\frac{1}{2}} \tag{5.11b}$$

It may be interesting to explore the significance of this quaternion in connection with the interaction potentials and associated force-distributions for specific pairs of interacting particles.

However, instead of going into these details here, we investigate alternative representation of the potential function based on the geometry of the system involved.

CHAPTER 5-B

A Geometrical Unified Representation of the Interaction Potentials

In this Chapter we attempt to obtain a unified representation of the four interaction potentials by means of a geometrical method. For this purpose, we notice that the four interaction potentials could be roughly represented as illustrated (not to scale) in Figure 5.1.

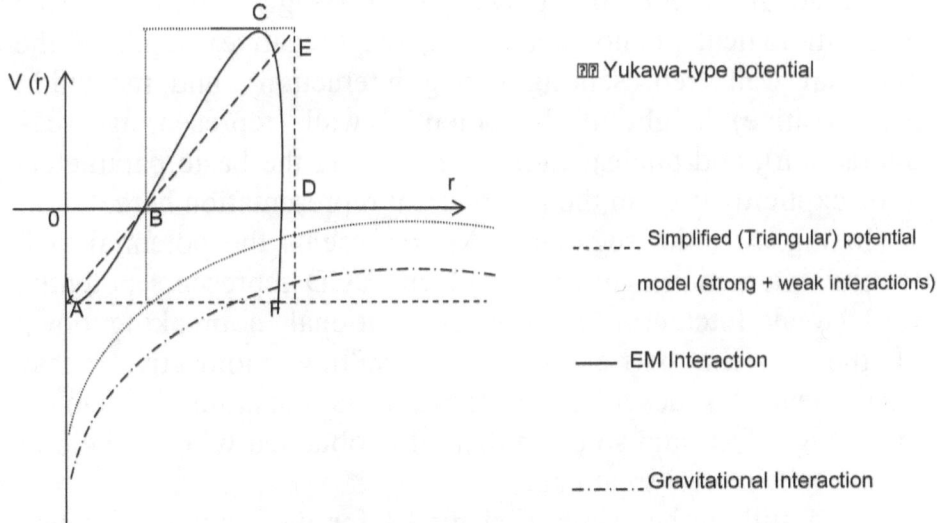

Figure 5.1.   A Rough Geometrical Representation of the Four Basic Interaction Potentials

5.8.   Strong and Weak Interaction

The strong interaction binds the quarks within a nucleon and the nucleons within a nucleus, the core at the center of an atom. For a neutron or a heavy nucleus, the statistical ensemble of quarks causes the potential function to be modified as shown in Figure 5.1. This familiar view gives rise to the customary concepts of the potential well (Yukawa-model), with depth proportional to the nuclear attractive force; and of a potential wall representing the repulsive force that causes nuclear decay. The decay of a heavy nucleus into lighter ones could be, in turn, accounted as the decay of a neutron (down-quark) into a proton (up-quark), an electron and an anti-electron-neutrino.

The strong force is operative within the nucleonic radius, while the weak force, within the nuclear radius. Within the atomic radius, the electromagnetic (EM) force sets in; while the gravitational force can be said to be relatively effective outside the domain of the atom. Of course, the EM and gravitational forces extend over the entire range (up to infinity), with the gravitational interaction playing increasingly dominant role in cosmological phenomena. Thus, the (negative) depth of the potential well (representing strong interaction), and its width; the (positive) height of the potential wall (representing weak interaction), and nuclear radius; are among the basic parameters to be explicitly used in the geometrical representation here.

In Fig. 5.1, the segment OAB represents the potential well (strong interaction), and the segment BCD represents potential wall (weak interaction.) As is conventional, a breaking down of the nuclear barrier and the resulting radioactive decay phenomenon is described in terms of a quantum mechanical tunneling effect, and solution thereof is obtained with the help of a wave mechanical treatment.

As a still further simplified model for the potential function of strong and weak interactions in combination, in Fig. 5.1, the triangular function described by $\Delta\ A\ E\ F$ (i.e., the combination of the region $\Delta\ A\ O\ B$ acting as the potential well for the strong interaction, and the region $\Delta\,B\,E\,D$ acting as the potential wall for the quantum tunneling or radioactivity for the weak interaction), may be considered together. The steep drop $E\,D\,F$ represents release of a higher level. The positive charge of the proton or a helium-ion (α-particle), for instance, and the negative charge of the electron (e^- or β-particle) released from the nuclear decay of the type

$$n \to P + e^- + v_e *$$
$$H_{e2} \to H_{e2}^+ + e^- + v_e *$$

cause the onset of the EM interaction. The EM interaction between charged particles as well as the gravitational interaction operating for all material particles and bodies gradually becomes relatively more significant as the electric charges and masses increase, even over long distances.

As discussed in Chpter 5-A, the strong interaction potential can be simply written as

$$V_1 = - (k/2)\, r^2 \tag{5.12}$$

where (-k) is the (negative, to signify attractive force) elastic constant for strong force (ECSF), and the suffix 1 designates the strong force.

Based on the ranges of the strong and weak interactions in this simple geometrical model, let us introduce the basic model parameters (fig. 5.1):

OA = V_0 = depth of the nuclear well binding the quarks
OB = γ_n = neutron radius
OD = γ_a = typical radius of the radioactive nucleus
DE = V_0' = height of the potential wall

Thus the coordinates of the points of interest are as follow

$$O \to (0,0)$$
$$A \to (0, -V_0)$$

$$B \rightarrow (\gamma_n, 0)$$
$$E \rightarrow (\gamma_a, V_0')$$

Using the equation for a straight line AE in the (x,y) plane, $y = mx + c$, here we have

$$m = \text{slope} = \frac{ED}{BD} = \frac{V_0'}{\gamma_a - \gamma_n}$$
$$c = y - \text{intercept} = -V_0$$

Therefore, we get the equation of the straight line AE representing the combination of the (remainder of the strong and) weak interaction as (designated by the suffix 2 below)

$$+V_1(\gamma) + V_2(\gamma) = \left(\frac{V_0'}{\gamma_a - \gamma_n}\right)\gamma - V_0$$

As the point B $(\gamma_n, 0)$ lies on this line, we obtain the relationship

$$\left(\frac{V_0'}{\gamma_a - \gamma_n}\right)\gamma_n - V_0 = 0$$

i.e.,

$$V_0' = \left(\frac{\gamma_a - \gamma_n}{\gamma_n}\right)V_0 = -\left(1 - \frac{\gamma_a}{\gamma_n}\right)V_0$$

Hence, the equation of the straight line AE becomes

$$V_1(\gamma) + V2(\gamma) = V_0\left(\frac{\gamma}{\gamma_n} - 1\right) \tag{5.13}$$

which is independent of the nuclear radius γ_a, depending only on the nucleonic radius.

5.9. The EM Interaction

The electromagnetic (EM) force and interaction potential are given by the Coulomb law:

$$F_3(r) = \frac{q_1 q_2}{4\pi r^2} = \frac{n_1 n_2 e^2}{4\pi r^2} = \frac{\partial V_3(r)}{\partial r}$$

i.e.,
$$V_3(r) = +\frac{n_1 n_2 e^2}{4\pi} \int \frac{1}{r^2} dr = -\frac{n_1 n_2 e^2}{4\pi r} \qquad (5.14)$$

where the suffix 3 designates EM interaction, r is the distance between the two electric charges $q_1 = n_1 e$ and $q_2 = n_2 e$, with n_1 and n_2 units of unit (electron) charge *(e)*, respectively; q_1 and q_2 being written with positive *(+)* or negative *(-)* charge based on the type of the charge(s) involved.

5.10. The Gravitational Interaction

Writing Newton's universal gravitational interaction between two masses m_1 and m_2 of a gravitational potential, $V_4(r)$, we have

$$F_4(r) = -g\frac{m_1 m_2}{r^2} = \frac{\partial V_r(r)}{\partial r}$$

i.e.,

$$V_4(r) = -gm_1 m_2 \int \frac{dr}{r^2} = \frac{Gm_1 m_2}{r} \qquad (5.15)$$

5.11. Composite Potential

The composite potential due to the four basic interactions could now be simply written by collecting the various contributions as (with slight change of notation for distinction)

$$V_T = \sum_{k=1}^{4} V_k(r)$$

$$V_T = \frac{-k}{2}\gamma^2 + V_0\left(\frac{\gamma}{\gamma_n} - 1\right)$$

$$-\frac{n_1 n_2 e^2}{4\pi\gamma} + \frac{Gm_1 m_2}{\gamma}$$

$$= -V_0 + \frac{V_0}{\gamma_n}\gamma + \frac{(-k)}{2}\gamma^2$$

$$-\left(\frac{n_1 n_2 e^2}{4\pi} - Gm_1 m_2\right)\frac{1}{\gamma}$$

$$= \sum_{k=-1}^{2} C_k \gamma^k$$

i.e.,   (5.16)

where the coefficients C_k, $(k = -1,0,1,2)$ are given as

$$C_{-1} = -\frac{n_1 n_2 e^2}{4\pi} + Gm_1 m_2$$

$$C_0 = -V_0$$

$$C_0 = -V_0$$

$$C_1 = \frac{V_0}{\gamma_n} = -\frac{C_0}{\gamma_n}$$

$$C_2 = \frac{-k}{2}$$

(5.17 a,b,c,d)

Of course, alternative representations of the strong and weak interactions in the form of rectangular or trigonometric function-type potential well and potential wall, respectively, are also possible. For instance, adopting the Yukawa model type representation, one could write

$$V_1 + V_2(\gamma) = \begin{cases} -V_0, 0 \leq \gamma \leq \gamma_n \\ +V_0', \gamma_n < \gamma \leq \gamma_a \\ 0, \gamma_a < \gamma \end{cases}$$

If, on the other hand, a trigonometric curve (solid line in Fig5.1) is employed, this alternative representation could be expressed as

$$V_3(\gamma) = V_0 \cos(\pi - \gamma)\delta(\gamma_n - \gamma) + \\ + V_0' \sin(\gamma - \gamma_n)\delta(\gamma_a - \gamma)$$

where the δ-function is defined as follows:

$$\delta(x) = \begin{cases} 1, x \geq 0 \\ 0, x < 0 \end{cases}$$

For any of these types of representations, combination of all the four interactions could thus be represented in a unified form as a power series in the radial distance (r) for further analysis. For instance, as mentioned above, the composite potential could be used in the radial part of the Schrodinger equation and solved using numerical methods as appropriate for the form of the potential function involved.

5.12. Solution of the Wave Equation with the Basic Interaction in the 3-D S-Space (γ, θ, Φ)

The Schrodinger wave equation in the spherical polar coordinates (γ, θ, Φ) for a spherically symmetric ("central") force-field, represented by a potential function $V(\gamma)$, is given as

$$\nabla^2 \psi + \frac{2M}{h^2}\left[E - V(\gamma)\right]\psi = 0$$

where ∇^2 is the Lapalacian operator given in polar spherical coordinates in the form

$$\nabla^2 = \frac{1}{\gamma^2}\frac{\partial}{\partial \gamma} + \frac{1}{\gamma^2 \sin \theta}\frac{\partial}{\partial \theta}\left(\sin \theta \frac{\partial}{\partial \theta} \right)$$

$$+ \frac{1}{\gamma^2 \sin^2 \upsilon}\frac{\partial^2}{\partial \phi^2}$$

$$(5.18a)$$

Here M is the reduced mass of the system and E the energy of the particle in question, and h is Planck constant. By the usual separation of variables t, γ, θ and ϕ, and by postulating the multiplicative component wave functions $\Phi(\phi)\oplus(\theta)R(\gamma)$, solutions of the wave equation could be written in the form

$$\psi (\gamma, \theta, \phi) = \frac{X(\gamma)}{\gamma}\sum_{l=i}^{\infty} \sum_{m=-l}^{l} Y_l^m (\theta, \phi)$$

$$(5.18b)$$

where the functions $Y_l^m (\theta,\phi)$ are the spherical harmonics of order (l,m) (see Eqn. 2.11b) and the radial function

$$X(\gamma) = \gamma R(\gamma)$$

satisfies the differential equation of the form

$$\frac{d^2 X}{d\gamma^2} + \frac{2M}{h^2}\left[E - V'(\gamma) \right]X = 0$$

In the above equation, $V'(\gamma)$ is the effective potential given as

$$V'(\gamma) = V_T(\gamma) + \frac{l(l+1)h^2}{2M\gamma^2}$$

$$= \sum_{k=-2}^{2} C_k \gamma^k$$

Here, the coefficient

$$C_{-2} = \frac{l(l+1)h^2}{2M}$$

and the remaining coefficients (C_{-1}, C_0, C_1, C_2) are as already given before (Eqns. 5.16 and 5.17 a,b,c,d).

The spherical harmonic functions are given as (see Eqn. 2.11c,d)

$$Y_l^m(\theta,\phi) = \theta_l^m(\theta)\Phi_m(\phi)$$

$$\theta_l^m(\theta) = \left[\frac{(2l+1)}{2}\frac{(l-m)!}{(l+m)!}\right]^{\frac{1}{2}} P_l^m(\cos\theta)$$

$$\Phi_m(\phi) = \frac{1}{\sqrt{2\pi}}e^{im\Phi}$$

(5.18c)

The radial equation could be solved analytically or numerically as appropriate. This leads to a complete solution in principle.

5.13. Solution in the 3-D T (Time-Space) $(t_\gamma, t_\theta, t_\phi)$

The formulation of the wave-equation in the spherically symmetric 3-D T time-space can be followed by a treatment similar to that employed for the 3-D S above. However, obviously, in this case only those systems are to be considered for which a potential function, $V(t_\gamma)$, can be devised which depend on the positive arrow of time, and those that are functions of 'angular time variables' t_θ and t_ϕ. Many "mysterious" physical phenomena, such as the dark energy, the dark matter, gravity wave, mixing and entanglement of the neutrinos and electrons, cosmological inflation, etc., may be processes involving these extra dimensions of time, essentially referring to motion in a plane perpendicular to the direction of observation, as already mentioned before.

Thus the wave-function in purely time-domain can be said to follow the wave equation similar to the Schrodinger equation, Eqn. (5.18), but where the spatial coordinates are replaced by the corresponding time-coordinates. In a spherically symmetric time-space, this replacement is based on the time-coordinates t_γ, t_θ, t_ϕ. The replacement

$$\gamma \rightarrow t_\gamma$$
$$\theta \rightarrow t_\theta$$
$$\phi \rightarrow t_\phi$$

then yields the 3-D T equations and solutions analogous to those for the spatial coordinates (γ, θ, Φ), Eqn. (5.18a)., i.e., of the type and form indicated below.

$$\overline{\psi}\left(t_\gamma, t_\theta, t_\phi\right) = \frac{\overline{X}\left(t_\gamma\right)}{t_\gamma} \sum_{l=0}^{\infty} \sum_{m=-l}^{l} \overline{Y}_l^m \left(t_\theta, t_\phi\right) \tag{5.19a}$$

where the radial time-function

$$\overline{X}\left(t_\gamma\right) = t_\gamma \overline{R}\left(t_\gamma\right) \tag{5.19b}$$

obeys the second order differential equation

$$\frac{d^2 \overline{X}}{dt_\gamma^2} + \frac{2\overline{M}}{h^2}\left[\overline{E} - \overline{V}'\left(t_\gamma\right)\right]\overline{X} = 0 \tag{5.20}$$

the barred quantity $\overline{\varepsilon}$ in the purely time-domain being analogous to the quantity $\underline{\varepsilon}$ in the 3-D S (spatial space) as defined and discussed in the preceding Sections. More specifically, the time-domain potential function, $\overline{V}'(t_\gamma)$, must be suitably devised to appropriately describe the phenomenon in question, based on phenomenological factors and pertinent basic interactions.

In passing, a special remark could be made here. It is natural to expect that the domain of the 3-D time-space (3-D T) may entail phenomena beyond the bounds of conventional physics.

In particular, such 'non-physical' phenomena may include brain and neuron-related activities and experiences. Examples of such non-conventional (from the standpoint of physics) could be human consciousness, memory, dreams, clairvoyance, psychic reactions and experiences (meditational or substance-induced), especial cognitive capabilities, and so on.

5.14. Solution in the 6-D ST (6-D Spacetime Continuum)

For a more rigorous treatment, the spatial interval γ, of the 3-D S as well as the 3-D T (conventional time-interval using the positive arrow of time) time-interval t_γ must be replaced by the equivalent space-time interval, in the 3-D ST (spacetime continuum).

In the 6-D ST, using the Cartesian coordinates, the value of Γ between two points $P(x_1, x_2, x_3, x_4, x_5, x_6)$ and $x_1', x_2', x_3', x_4', x_5', x_6'$ is given by

$$\Gamma = \left[\left(x_1' - x_1\right)^2 + \left(x_2' - x_2\right)^2 + \left(x_3' - x_3\right)^2 + \left(x_4' - x_4\right)^2 + \left(x_5' - x_5\right)^2 + \left(x_6' - x_6\right)^2 \right]^{1/2}$$

$$= \left[\sum_{k=1}^{6} \left(x_k' - x_k\right)^2 \right]^{1/2}$$

For simplicity, choosing the origin of the 6-D ST orthogonal hyperspace at the point P, one can write

$$\Gamma = \left[\sum_{k=1}^{6} \left(x_k'\right)^2 \right]^{1/2}$$

Using the spherical polar coordinates, the value of this 6-D interval becomes

$$\Gamma = \left[\gamma^2 - \left(ct_\gamma\right)^2 \right]^{1/2}$$

which is simply the distance of the point P' from the origin. The variable T must be used in place of either r or t_γ for a complete treatment of a physical and neo-physical system in the 6-D ST hyperspace. Alternative analogous angular variables in the 6-D ST hyperspace, say, coordinates (ε, G), could also be introduced to replace the 3-D S coordinates (θ, ϕ) or the 3-D T coordinates (t_θ, t_ϕ). Full 6-D ST hyperspace solutions could then be developed in analogy to the corresponding equations and solutions of the 3-D S or 3-D T spaces. These equations and solutions are expected to provide a comprehensive and all-inclusive description of the elementary particles and the four basic interactions, and, as such, could be referred to as one form of the Theory of Everything (TOE). This 6-D ST-based approach, based on the `spring theory' of the strong and weak interactions, can be said to provide a form of the universal theory of elementary particles and interactions (UTOEPI).

CHAPTER 6

Special Features of the Three-Dimensional Time or Six-Dimensional Spacetime (6-D St)

"All the conditions of happiness are realized in the life of the man of science."

Bertrand Arthur William Russell (1872-1970)

6.1. Introduction

This Chapter presents further arguments to highlight the potential applications and justifications for extending the concept of 4-D ST to a new paradigm of the six-dimensional space-time (6-D ST) continuum by introducing a three-dimensional time-space (3-D T) orthogonal to the 3-D S, with the three time co-ordinates also forming a right-hand orthogonal Cartesian reference frame as the 3-D S. Apart from the conventional time (associated with the "positive arrow" of time of increasing entropy and of common physical phenomena and every-day experience), the additional two orthogonal time dimensions may be regarded analogous to the familiar Kaluza-Klein ("*curled-up*") variables playing significant role only in super-high energy (Planck scale) range pertinent in sub-nuclear particles on the one hand and in cosmology, on the other.

6.2. Examples of Potential Implications

The implications of treatment of (high-energy) physical events in this 6-D ST hyperspace are briefly overviewed and recapitulated. In passing, especial implication in terms of potential theoretical explanations of several (high-energy) physical phenomena in the realms of elementary particle physics and cosmology, including a few already dealt with earlier in this book briefly, are summarily mentioned. These include, for instance,

(i)   A simple coherent and unified model for the elementary particles (the Standard Model) and the four basic interactions: strong, weak, electromagnetic (EM) and gravitational;

(ii)   The neutrino entanglement: Apparent transformation of the electron neutrino (y_e) into muon neutrino (y_μ) or tau neutrino (y_τ) (or, manifestations of these three types of neutrinos as three quantum states of one basic particle) as rotations in the extra time dimensions;

(iii)   Gravity wave: Propagation of the gravity wave could entail the extra dimension(s) of time, making it difficult to detect or observe the same through conventional (single-time dimensional) means or approach;

(iv)   Dark Energy and Dark matter: The likely presence of the dark energy (in the form of electromagnetic or matter-wave) with polarization in the additional time co-ordinates or planes, making it difficult to detect or observe the same through conventional means;

(v)   Post-big bang Inflation: The Inflation process in the early universe immediately after the big bang could have been affected or originated under a sudden large expansion in the 6-D ST;

(vi)   A flat-universe approximation: Introduction of space-time continuum curvature, causing a relatively more artificial interpretation of mass (as a result of such a curvature, or causing it) is not implied or necessary; a flat 6-D ST geometry could be naturally considered acceptable and sufficient;

(vii) Inherent spherical symmetry: Mathematical models based on the naturally expected spherical symmetry in all (6-D) space and time co-ordinates (i.e., a complete space-time symmetry) are easily obtainable. In particular, a mutual equivalence and potentiality of an inter-conversion of the spatial and time 'spaces' becomes more transparent and quantifiable imbibing consistency in the relativity and quantum theories, as a simple corollary.

Many additional implications of importance to particle physics and cosmology are likely to emerge from the above type of extension in the representation and description of the time space (3-D T and 6-D ST). As is customary in the conventional string theory, the extra time co-ordinates may be considered to be dormant *("curled")* in the normal, macroscopic world, but to play a vital role in the case of a microscopic and quantum (Planck-scale, sub-atomic, sub-nuclear) as well as of cosmological systems, as mentioned above.

6.3.   Inter-conversion of Space and Time in the 6-D ST

Just as the conventional quantum uncertainty relations for energy (E) and time (t); and for the position (x) and momentum (p), expanded to apply in the 6-D ST,

$$\Delta E_j \, \Delta t_j \; \sim \; \hbar \; \sim \; \Delta x_j \, \Delta p_j \; , \qquad (6.1a)$$

i.e.,, $\Delta E_j \quad \sim \; \hbar / \Delta t_j \qquad\qquad\qquad (6.1b)$

$\Delta x_j \quad \sim \; \Delta E_j \, \Delta t_j / \Delta p_j \; \sim \; c \, \Delta t_j \; , \qquad (6.1c)$

(j=1,2,3)

where we have used the relativistic relation $\Delta E_j = \Delta p_j c$ (see Section 2.16.4). Eqn. (6.1b,c) indicate how a very minute uncertainty in any one or more of the time coordinates could be associated with a very large uncertainty in either energy component(s) and/or the spatial coordinate(s). Such changes or uncertainties, in turn, could lead to the phenomena of the dark energy on the one hand and of the inflation-process immediately following the big bang, on the other.

More specifically, the above correlation in uncertainty in (the conventional) time and the corresponding uncertainty in energy with the Planck constant ($\hbar$), can also be extended to the orthogonal time components bearing similar relationships to the additional energy components; and this could be potential sources of unobservable energy of huge proportions (for miniscule uncertainties in orthogonal time components; note that, in principle, for $\Delta t_j \rightarrow 0$, we get $\Delta E_j \rightarrow \infty$ and $\Delta x_j \rightarrow \infty$.) In particular, in the spherical polar co-ordinate system, a large uncertainty in Δt_r may still incorporate arbitrarily small values in the 'angular' time-coordinates, Δt_θ and/or Δt_ϕ, thereby leading to the possibility of cosmologically enormous energy (polarized) in the longitudinal and/or latitudinal directions of the three-dimensional time-space (3-D T.) For instance, for $\Delta t_j \sim 10^{-43}$ sec. (Planck time), a 10^{16} order of magnitude enhancement in energy (and $\Delta t_j \times c = 10^{-33}$ cm, i.e., for an expansion over a Planck distance) could be expected. This type of phenomena could possibly form, or contribute to, the explanation of 'mysterious' cosmological processes such as inflation immediately following the big bang, the dark energy, and dark matter, and quantum gravity (gravitons carrying the energy in quanta or waves polarized in the orthogonal time dimensions.)

6.4. Additional Potential Applications

As mentioned, the extra dimensions of time could be considered as hidden or *'curled up'* (to borrow a phrase from the 'string theory') due to our inability to experience their effect directly, in common measurements or day-to-day experience. However, their contributions in the structure and interactions of elementary particles, i.e. in the ultra-high energy range (as in the case of the big bang and inflation phenomena) as well as in astrophysical and cosmological events (that is, in the ultra-microscopic phenomena, such as the big bang, and in the ultra-macroscopic systems such as galactic acceleration) could be quite significant. For instance, in the cases of the dark energy or the gravity waves, the pertinent waves could be polarized in these extra time dimensions and hence undetectable by conventional means which are able to detect only waves polarized in the conventional time dimension. This type of effects can be characterized as speculative at this stage, but cannot be ruled out altogether; and further analysis in this direction may prove to be useful.

The mathematical simplicity inherent in this approach for the description of relevant physics is self-evident and, apart from ease in treatment and analysis, may be closer to the physical reality by virtue of a larger degree of symmetry between the space and time co-ordinates. In fact, introduction of a 3-D T is inevitable for consistency even within the relativity theory, as discussed in Chapter 1.

A large number of enigmatic and perplexing processes and observations in particle physics and cosmology could be correlated to, or explained by, the assumption of extra orthogonal time co-ordinates, as already briefly alluded to above. Additional features which could perhaps be related to the conjecture of three dimensional time space (3-D T) might include the following:

(a) The ordinarily observable ('visible') matter appears to be roughly less than 5% of total matter and energy (including the dark matter or dark energy) in the universe. This could be the result of transverse coordinates (including the two time co-ordinates, t_θ and t_ϕ) being 'invisible' and only one co-ordinate, t_r, lending itself to direct visualization in ordinary observations on the cosmic scale.

(b) The quarks appear or disappear only in pairs, possibly because the origin and decay of quarks take place only in the two orthogonal time-space co-ordinates t_θ and t_ϕ, which, in turn, could be due to conservation of angular momentum along the corresponding directions, similar to the case of the neutrino carrying the angular momentum in the neutron decay (as was originally postulated about the existence of the neutrino.)

(c) The (real) spatial co-ordinate, r, and the three (imaginary-type) time co-ordinates, ict_r, ict_θ, and ict_ϕ, regarded as non-commutative axes of rotation in the 3-D T, could be considered as constituting a quaternion set, and this fact could be utilized to construct a simple model for a Unified Theory of Elementary Particles and Interactions (UTOEPI) including the Higgs boson (the God Particle.)

Thus, at the risk of repetition, it can be noted that for a reconciliation of the concepts of relativistic invariance and quantum uncertainty, especially pertinent to the microscopic world and phenomena, and/or a simple interpretation of fundamental physical entities (particles and fields as well as cosmological evolutionary events), the introduction of time co-ordinates orthogonal to the direction of observation is practically indispensable. This is tantamount to treating time as a 3-D vector analogous to, and orthogonal to, the conventional 3-D S (spatial co-ordinates.) Thus, we are naturally led to the concept of a 6-D space-time continuum for developing a self-consistent unified theory of elementary particles and the basic interactions—the so-called UTOEPI referred to above. Such a theory stands a better chance of successfully describing the universe from the big bang to the evolution of large scale features and processes of the universe. In such a treatment or model-formulation, the artifice of space-time curvature is not required, and a flat-universe assumption with more simplistic presentation of space and time variables in terms of ordinary Cartesian co-ordinates should suffice. At the same time, such an approach could help model in a rather trivial fashion, as well as more realistically and accurately, the theoretical implications to explain and resolve many of the puzzles and conflicting results in the realms of cosmology and particle physics. Of course, more investigation is required to confirm the speculative nature of these assertions, but the existence of extra, orthogonal time components could open the door to enormously significant results *('new physics')* in many diverse areas, especially in particle physics and cosmology. If such a hypothesis were to be accepted as valid, mathematical complexity or ambiguity in alternative models, such as the string theory, the bane theory, the M-theory, the superstring theory, etc., would then become mute.

The primary implications of this 6-D ST-based model can be best viewed in the framework of the quaternion algebra which represents rotation of a vector, or equivalently, of the coordinate axes, in a 4-D complex orthogonal space constituted of *one real* (radial separation in 3-D S) and *three imaginary* (3-D T) dimensions. The set of successive rotations can be easily shown to correspond to introduction of the set of the four basic interactions. The model furnishes a simple, natural explanation of the types and number of elementary particles in the Standard Model.

Furthermore, a simple relationship between the particle mass and its conserved interaction parameters is obtained. As an important corollary, it can be said that the particle mass is a universal consequence of its interaction with the fields of interaction, the square of each interaction constant adding, in a near-linear fashion (with one exception which leads to universal evolution of matter in the universe) to the mass of the particle. Last, but not the least, this treatment leads to scopes for obvious generalizations of the model presented here (UTEOPI) to include, if and as necessary, a self-consistent description (or prediction) of phenomena that might become observable in the future, at energy ranges of significantly higher orders of magnitude than presently available. The proposed and newly observed Higgs boson (the 'God-Particle') by the LHC[4] furnishes a especial example of such a generalization. Finally, this model also leads to a simple explanation of the nuclear beta-decay and its important implications for the cosmological evolutionary transformations in the universe, from quarks to atoms and molecules of increasing complexity, and further to biological systems, DNA, and finally leading to life and ourselves.

6.5.   The Case of 6-D Spherically Symmetric System

To further illustrate the advantages of the 6-D ST based description and modeling, a review and recapitulation of the vector and quaternion algebra is first presented here for convenience of reference, followed by consideration of the case of a 6-D spherical symmetric system and certain implications thereof.

6.5.1. Vector Representation

For a spherically symmetric system, the 6-D spacetime can be assumed to be describable, for simplicity, in terms of the single spatial coordinate (the radius vector r), plus the three orthogonal coordinate axes representing 3-D time dimensions (t_1, t_2, t_3 or t_r, t_θ, t_φ, respectively), treated as the time-vector and abbreviated here as 3-D T. The remaining two spatial dimensions (θ and φ, in the spherical polar coordinate system) are omitted here, since the particle or system behavior is assumed to be independent of the latitudinal (polar angle θ) and longitudinal (azimuthal angle φ) coordinates, depending only on (the relative separation or distance) r. Hence, in such a spherically symmetric system, consideration in terms of a 4-D coordinate system (r, t_r, t_θ, t_φ) suffices, which is a reduced form of the full 6-D spacetime (6-D ST).

The vector representation of the 4-D spacetime is based on the orthonormal vectors ($\hat{i}, \hat{j}, \hat{k}$) applied to the three orthogonal time coordinates (t_1, t_2, t_3) in combination with the unit vector $\hat{r}$

(along r) normal to the 3-D T. A vector $\vec{V}$ in this 4-D ST can therefore be represented in the form:

$$\vec{V} = r\hat{r} + x_4\hat{i} + x_5\hat{j} + x_6\hat{k} \tag{6.2}$$

where
$$x_1 = ict_1, \, l = 4,5,6$$

c being the velocity of light,

$$i = \sqrt{-1}$$

and
$$\left(\hat{r}\right)^2 = \left(\hat{i}\right)^2 = \left(\hat{j}\right)^2 = \left(\hat{k}\right)^2 = 1$$

$$r \cdot \hat{i} = r \cdot \hat{j} = r \cdot \hat{k} = \hat{i} \cdot \hat{j} = \hat{j} \cdot \hat{k} = \hat{k} \cdot \hat{i} = 0$$

and the scalar product, represented by dot, is commutative.

6.5.2. Quaternion Representation

Here, we invoke a geometric interpretation of a general quaternion, Q, defined as

$$Q = \alpha_0 + \alpha_1 i + \alpha_2 j + \alpha_3 k \tag{6.3a}$$

where α_0, α_1, α_2, α_3 are real and the quaternion generators (i,j,k)—not to be confused with the orthonormal vectors $\left(\hat{i},\hat{j},\hat{k}\right)$ —obey the defining, non-commutative, algebra (see Section 2.8):

$$i^2 = j^2 = k^2 = -1 \tag{6.3b}$$

$$ij = -ji = k; \tag{6.3c}$$

$$jk = -kj = i; \tag{6.3d}$$

$$ki = -ik = j \tag{6.3e}$$

and, hence, $ijk = -1$ (6.3f)

The suggested geometrical interpretation is simply based on the assumption that the generator i is equivalent to a rotation of the vector, $\vec{r} = r\hat{r}, |\vec{r}| = r = \alpha_0$, say, by $\frac{\pi}{2}$ radians, in the anti-clockwise sense, to coincide with the direction of the vector $\hat{i}$. This process forms the complex space $(\hat{r}, \hat{i})$.

Similarly, geometrically, the quaternion generators j and k can be interpreted as additional rotations by $\frac{\pi}{2}$ (of $\hat{i}$ to coincide with $\hat{j}$; and then another rotation of $\hat{j}$ to coincide with $\hat{k}$), so as to result in the 4-D ST, with the first spatial coordinate r along $\vec{r}$ and the three time-coordinates ($x_4 = ict_1$, $x_5 = jct_2$, $x_6 = kct_3$) together forming a four-dimensional orthogonal hyperspace satisfying the quaternion algebra as well as the vector algebra as mentioned above.

6.5.3. Vectors in the 6-D Spacetime

Any vector $\overrightarrow{OP}$ (with magnitude OP) in the $(\hat{r}, \hat{i})$ space could be represented in terms a complex variable z

$$z = [(OP)\cos\beta] + i[(OP)\sin\beta], \ i = \sqrt{-1} \qquad (6.4a)$$

$$|z| = \{[(OP)\cos\beta]^2 + [(OP)\sin\beta]^2\}^{\frac{1}{2}}, \qquad (6.4b)$$

Similarly, in the 4-D ST hyperspace defined by the quaternion Q, (see Eqn. 2.4e), and the magnitude of a Q, analogous to the magnitude of 4-D vector in this hyperspace, is defined by

$$|Q| = \overline{Q} = \pm\left(\alpha_0^2 + \alpha_1^2 + \alpha_2^2 + \alpha_3^2\right)^{\frac{1}{2}} \qquad (6.4c)$$

where the sign has been explicitly included on the right for the sake of generality, since clearly both signs are permissible in taking the square-root of any quantity.

6.6. Application To The Elementary Particle System

We now apply to the case of the system of elementary particles the quaternion algebra and the associated geometrical interpretation in terms of rotations in the 4-D hyperspace $(\hat{r}, \hat{i}, \hat{j}, \hat{k})$, with a (real) spatial coordinate (the radial distance r) and the three (imaginary) time coordinates $((ict_1, jct_2, kct_3)$ or $(x_4, x_5, x_6))$. We note at the outset that, as necessary, a direct correspondence of r with α_0, and of (i,j,k) with $(\alpha_1, \alpha_2, \alpha_3)$, respectively, can be obviously invoked (see Eqn. 2.4e).

First, we notice that for each family (F_1, F_2, F_3) of elementary particles, the masses of the four family members are different, mutually (intra-family) as well as with respect to the masses of the corresponding members of other families (inter-family.) However, the spin-values and the values of the charges (S, w, q for strong, weak and EM interactions, respectively) of the first member of each family are the same, viz.,

$$(s, S, w, q) = (1/2, 1, 1/2, 2/3) \qquad (6.5a)$$

Similarly, the spin-value and the charges of the second member of each Family are identical; the values of the third member of each family are identical; and the values of the fourth member of each Family are identical.

Based on this observation, it is useful to define the following general 'elementary particle quaternion (EPQ)'

$$Q = \alpha_0 s + \alpha_1 S i + \alpha_2 w j + \alpha_3 q k, \qquad (6.5b)$$

where $(\alpha_0, \alpha_1, \alpha_2, \alpha_3)$ are real coefficients to the spin-value (s), the strong charge (S), the weak charge (w), and the EM charge (q), respectively. The magnitude of this quaternion then becomes

$$\overline{Q} = \pm \left[(\alpha_0 s)^2 + (\alpha_1 S)^2 + (\alpha_2 w)^2 + (\alpha_3 q)^2 \right]^{\frac{1}{2}} \qquad (6.5c)$$

We can generally explore the possibility of correlating the properties of the particles with the characteristics of the elementary particle quaternion (EPQ).

6.6.1. Simple Model of the Particle Mass: Case 1(A) for s = 1/2:

As an example, we consider derivation of a simple model of the particle mass. For this, we start with the simplest representation of the EPQ in the form

$$Q = s + Si + wj + qk \tag{6.6a}$$

with magnitude

$$\overline{Q} = \pm\left[s^2 + S^2 + w^2 + q^2 \right]^{\frac{1}{2}} \tag{6.6b}$$

Specifically, for the first, the second, the third, and the fourth member of each family of elementary particles, we introduce the quaternions Q_1, Q_2, Q_3, Q_4, respectively.

$$Q_1 = \left(\frac{1}{2}\right) + (1)i + \left(\frac{1}{2}\right)j + \left(\frac{2}{3}\right)k = \frac{1}{2} + i + \frac{1}{2}j + \frac{2}{3}k$$

$$Q_2 = \left(\frac{1}{2}\right) + (1)i + \left(\frac{1}{2}\right)j + \left(-\frac{1}{3}\right)k = \frac{1}{2} + i + \frac{1}{2}j - \frac{1}{3}k$$

$$Q_3 = \left(\frac{1}{2}\right) + (0)i + \left(-\frac{1}{2}\right)j + (-1)k = \frac{1}{2} - \frac{1}{2}j - k$$

$$Q_4 = \left(\frac{1}{2}\right) + (0)i + \left(-\frac{1}{2}\right)j + (0)k = \frac{1}{2} - \frac{1}{2}j$$

with magnitudes $\overline{Q}_1, \overline{Q}_2, \overline{Q}_3, \overline{Q}_4$

$$\overline{Q}_1 = \left[\left(\frac{1}{2}\right)^2 + (1)^2 + \left(\frac{1}{2}\right)^2 + \left(\frac{2}{3}\right)^2\right]^{\frac{1}{2}} = \sqrt{\frac{35}{18}}$$

$$\overline{Q}_2 = \left[\left(\frac{1}{2}\right)^2 + (1)^2 + \left(\frac{1}{2}\right)^2 + \left(-\frac{1}{3}\right)^2\right]^{\frac{1}{2}} = \sqrt{\frac{29}{18}}$$

$$\overline{Q}_3 = \left[\left(\frac{1}{2}\right)^2 + (0)^2 + \left(-\frac{1}{2}\right)^2 + (-1)^2\right]^{\frac{1}{2}} = \sqrt{\frac{3}{2}}$$

$$\overline{Q}_4 = \left[\left(\frac{1}{2}\right)^2 + (0)^2 + \left(-\frac{1}{2}\right)^2 + (0)^2\right]^{\frac{1}{2}} = \sqrt{\frac{1}{2}}$$

An alternative option is to consider $(S + w)$ as a single parameter S_w jointly, since S and w vary as a pair of parameters in the same manner. It is of interest to also examine the quaternion magnitudes if the spin-values are hypothetically assumed to be 0 or 1, instead of the usual (fermion) value $s = \frac{1}{2}$. Denoting the values as $\overline{Q}_i'$ corresponding to $s = 0$, and as $\overline{Q}_i''$ corresponding to $s = 1$, ($i = 1,2,3,4$), we obtain the values of the magnitudes of the corresponding EPQs. Note that the values $s = 0$ or $s = 1$ correspond to the cases of bosons, to be expected under the hypothesis of supersymmetry. We compute below the magnitudes of these additional EPQs, which are distinguished by labeling them as Case 1(B) and case 1(C), in contrast to the above ($s = \frac{1}{2}$ for fermions) case which is labeled as Case 1(A).

6.6.3. Case 1(B) for s = 0

$$\overline{Q}_1' = \sqrt{\frac{61}{36}}$$

$$\overline{Q}_2{}' = \sqrt{\frac{49}{36}}$$

$$\overline{Q}_3{}' = \frac{\sqrt{5}}{2}$$

$$\overline{Q}_4{}' = \frac{1}{2}$$

6.6.4. Case 1(C) for s = 1

$$\overline{Q}_1{}'' = \sqrt{\frac{97}{36}}$$

$$\overline{Q}_2{}'' = \sqrt{\frac{85}{36}}$$

$$\overline{Q}_3{}'' = \sqrt{\left(\frac{3}{2}\right)}$$

$$\overline{Q}_4{}'' = \frac{\sqrt{5}}{2}$$

The ± signs have been omitted in the above results for brevity. Note that Figure 2.3 is based on abscissa values corresponding to Case 1B above, but by omitting for simplicity the common spin value (s =1/2), i.e., effectively setting s = 0. Inclusion of the fermion spin value (s = ½) is tantamount to shifting all the three straight lines corresponding to Family 1, Family 2, and Family 3 (but not the straight line for the Higgs bosons) horizontally by $(1/2)^2 = ¼ = 9/36$ units. This shift does not alter the basic linear characteristic or the the numerical values of the particle masses, however.

It also must be noted that, just as in a 2-D complex- (or vector-) plane with real and imaginary components u and v, for the complex quantity $z = u + iv$, the magnitude $|z| = (u^2 + v^2)^{½}$

corresponds to the length of the vector $\vec{r} = \vec{u} + \vec{v}$, consistent with the Pythagorean theorem, so also the magnitude of a quaternion can be regarded as the square of length of the corresponding quaternion in the 4-D hyperspace ("hyper-vector"), as a generalization of the Pythagorean theorem for the hyperspace. With appropriate values in the present case these components correspond to the squares of the four conserved parameters, viz. s (spin), S (strong charge), w (weak charge), and q (*EM* charge) values. Moreover, for a more general case, these basic parameter values could be employed with suitable coefficients or weights, α_1, α_2, and α_3, instead of being taken with equal unit coefficient ($\alpha_0 = \alpha_1 = \alpha_2 = \alpha_3 = 1$), as assumed above.

Finally, it may be noted that in the present context, a positive (+) value of the particle quaternion is analogous to the length of the corresponding hyper-vector in the 'forward' direction along the positive arrow of time, i.e., in the direction of increasing entropy; whereas a negative (-) value, to the equal length of an oppositely directed hyper-vector, in the direction of decreasing entropy. The last case, though disallowed by the Third Law of Thermodynamics, may have some significance in exceptional cosmological phenomena, such as the black hole.

In a particle accelerator (or in a cosmological case such as the big bang), where the elementary particles are produced with extremely high energies traveling in all different directions, the special relativity theoretic assumption of a fixed and well-defined relative velocity along a specific single direction is no longer tenable. In other words, the concept of nonzero velocity components along two or three orthogonal coordinate axes must be permitted. This automatically implies a 3-D time, as indicated before (Chapter 1). This, in turn, leads to a system description in terms of the quaternion. Of course, the particle detectors presents only a planar projections of the 6-D ST, and this discrepancy in the actual system and its observed (planar projection) representation could well be the source of a degree of confusion in the interpretation of the reality or data regarding the world of the elementary particles, including the impact of the 'unphysical' negative entropy mentioned above. In particular, a holographic type of representation or interpretation of the physical reality is likely obtained in such a planar projection of the 6-D ST event, as has been speculated by some physicists.

6.6.5. Connection to the Magic Numbers

It is particularly interesting to also note the following results:

$$\pm \overline{Q}_3{}' - \overline{Q}_4{}' = \frac{\pm\sqrt{5}-1}{2} = \phi_1 \text{ or } \phi_2 \tag{6.7}$$

where

$$\phi_1\left(=\frac{\sqrt{5}-1}{2}\right) \tag{6.8a}$$

and

$$\phi_2\left(=\frac{-\sqrt{5}-1}{2}\right) \tag{6.8b}$$

are the well-known magic ratios (also called the magic numbers or the golden ratios) which embody many mathematically intriguing properties. These mathematical properties, in turn, could be potentially related to physical properties of the corresponding elementary particles, including the bosons (photons and Higgs bosons.)

In particular, the relation

$$\phi_1 + \phi_2 = \phi_1\phi_2 = -1 = i^2, \tag{6.8c}$$

a basic property of the magic numbers, could interchange the identity of different types of particles (entanglement phenomena.) By exploiting the interrelations between the magic numbers, useful relationships among the properties of the elementary particles could also be developed.

6.6.6. A Novel Representation of the Elementary Particle Parameters

It may be noted that the variation of the strong interaction parameter (S) and of the weak interaction parameter (w) for elementary particles in the Standard model follows a fixed pattern; viz., S=1 is associated with w= 1/2 (for strong interaction, involving quarks), while S=0 is associated with w= -1/2 (electron, muan, tau, and corresponding neutrinos.) Thus it is useful to introduce a new interaction parameter combining S and w in a suitable manner. We define the combination interaction parameter, Sw, as follows:

$$Sw^2 = \pm (S^2 + w^2)^{1/2}, \tag{6.9a}$$

The $\pm$ signs have been explicitly included above in the square-root to illustrate the natural way in which the magic numbers become an integral part of the interaction parameter presentation provided below.

In order to facilitate the introduction of the magic numbers and also to develop a new type of diagrammatical representation of the Families of the elementary particles, let us further introduce the parameter P defined as

$$P = Cos (\pi q) \tag{6.9b}$$

where q is the EM interaction parameter (electric charge) of the particle.

Table 6.1 now presents the complete set of interaction parameters and also certain derived parameters, viz.,

$$S_{wp} = S_w + P \tag{6.9c}$$

It is seen from Table (6.1) that the magic numbers or magic ratios (Φ_1, Φ_2) occur naturally if the parameter S_{wp} is introduced in the characteristics of the elementary particles. Again, the properties of the magic ratios, viz.,

$$\Phi_1 + \Phi_2 = \Phi_1\Phi_2 = -1,$$

could be used to derive many especial properties of the pertinent particles, and interrelations among them.

Table 6.1 The Interaction Parameters of the Elementary Particles and Connection with the Magic Numbers (Ratios) Φ_1 and Φ_2.

GROUP	Elementary PARTICLE	Strong Charge S	Weak Charge W	EM Charge q	$S_w =$ $(S^2+w^2)^{1/2}$	$P =$ $\cos(\pi q)$	$S_{wp} =$ $S_w + P$	Magic Ratio Φ_1,Φ_2	Higgs H
G1	Q_u, Q_c, Q_t	1	½	⅔	$\pm \sqrt{5}/2$	-½	$\pm\dfrac{\sqrt{5}-1}{2}$	Φ_1,Φ_2	0
G2	Q_d, Q_s, Q_b	1	½	-⅓	$\pm \sqrt{5}/2$	+½	$\pm\dfrac{\sqrt{5}+1}{2}$	$-\Phi_1$	0
G3	e^-, μ^-, τ^-	0	-½	-1	$\pm$ ½	1		¾, ½	0
G4	ν_e, ν_μ, ν_τ	0	-½	0	$\pm$ ½	1		¾, ½	0
Higgs Bosons	H_1^0, H^0, H_2^0	1	0	0					1

A diagrammatic representation of the elementary particles of the Standard model is shown in Figure 6.1. The symmetrical pattern of Fig. 6.1 points to the underlying symmetry in the system of the elementary particles, and also the elegance of the present model. Furthermore, the vacancy in the top-left segment of Fig. 6.1 (i.e., the angular region between the axis for q = 2/3 (or P = -1/2) and that for q = -1 (P = -1) seems to indicate a possibility of additional particles there. The Higgs particles belong in a different plane (for H = 1) than the familiar particles of the three Families (F1, F2, F3), which all correspond to H = 0.

A remark is also due concerning the question of supersymmetry. The present model remains unchanged if we were to reverse the electric charges of any or all of the particles, since the mass-formula is based only on the q^2 -value, or the diagrammatic representation is based on Cos (πq)-value, both of these factors remaining unchanged if q is replaced by -q.

Consequently, the present model could be extended to include a description of supersymmetry as well, should such a description be warranted on the basis of future theoretical and / or experimental investigations.

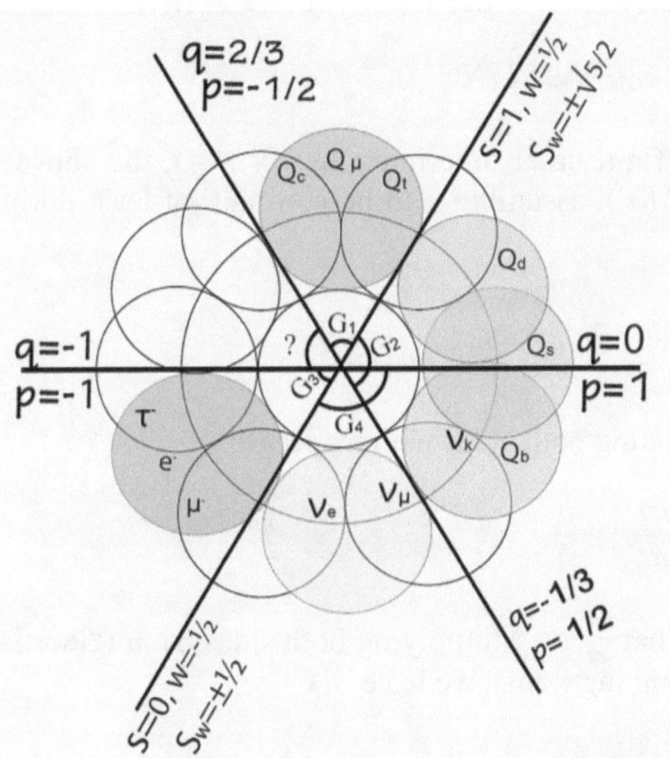

Figure 6.1.   Geometrical or diagrammatical representation of the Standard model elementary particles based on the derived interaction parameters Sw and P.

6.7. Relation between Classical and Relativistic Mechanics

We start with the basic invariance relation between the coordinates (x,y,z,t) of an inertial frame R with respect to those (x',y',z',t') of R'; R' moving with respect to R with a (relative) velocity v:

$$x'^2 + y'^2 + z'^2 + (ict')^2 = x^2 + y^2 + z^2 + (ict)^2$$

i.e., $r'^2 - c^2t'^2 = r^2 - c^2t^2$

For infinitesimal time interval $dt = t' - t$, the above becomes $d(r^2) = c^2 d(t^2)$, assuming c to be constant (at least during dt); so that,

$$2r dr = 2c^2 t dt, \text{ or } r\left(\frac{dr}{dt}\right) = c^2 t$$

Differentiating both sides with respect to t,

$$r\frac{d^2r}{dt^2} + \left(\frac{dr}{dt}\right)^2 = c^2$$

(6.10a)

Note that $\frac{dr}{dt} = v$. Multiplying both sides by m ("inertial mass") and rearranging terms, we have

$$mc^2\left(1 - \frac{v^2}{c^2}\right) = rm\frac{d^2r}{dt^2} = rmf \quad ,$$

where $f = \frac{d^2r}{dt^2}$ is the acceleration.

Using now Newton's Second law of motion for the applied force $F (= mf)$, we get

$$\left(\frac{F}{1-\frac{v^2}{c^2}}\right) r = mc^2 = E$$

,

where $E=$ is the relativistic energy.

i.e., $E = mc^2 = F_v r$ (6.10b)

where

$$F_v = \frac{F}{1-\frac{v^2}{c^2}}$$

(6.10c)

is the 'effective' (relativistic) force; and $F_v = F$ for $v << c$.

In other words, the relativistic energy (or work performed, and, hence, potential energy stored by the external force into the particle of mass m) is equal to the 'effective' force (F_v) multiplied by the distance (r), as expected.

The above is a direct connection between the classical (Newtonian) mechanics and the familiar relativistic energy formula, and also yields the rule for transforming the classical force into the relativistic one.

6.8. The Higgs Field and Special Relativity

Consider the relativistic equation for energy:

$$E^2 = p^2c^2 + m^2c^4$$

$$= m^2v^2c^2 + m^2c^4 \qquad (6.11a)$$

Can we relate this equation for the formula for the (hypothetical) Higgs field?

For this, assume

$$m = i\mu, \cdot i = \sqrt{-1},$$

Then, $E^2 = -\mu^2 v^2 c^2 - \mu^2 c^4$ (6.11b)

A plot of E^2 against m ^{2} would exhibit a double-well curve, the two wells on either side of the origin, with the origin also lying on the curve—characteristics expected of the Higgs field.

6.9.   Total Energy of a Particle

Integrating the kinetic energy $\epsilon = \frac{1}{2}mv^2$ of a particle of (instantaneous) mass m, rest mass m_o, and velocity v, starting from the rest state ($v=0$) to the highest possible limit ($v=c$), we have the total (available) energy, E, given as

$$\left(\because m = \frac{m_0}{\sqrt{1 - \dfrac{v^2}{c^2}}} \right).$$

$$E = \int_0^c \left(\frac{1}{2}mv^2 \right) dv = \frac{1}{2}m_o \int_0^c \frac{v^2}{\sqrt{1 - \dfrac{v^2}{c^2}}} dv = \frac{1}{2}m_o c^2 \int_0^c \frac{\left(v^2 \div c^2 \right)}{\sqrt{1 - \dfrac{v^2}{c^2}}} dv = \left(m_o c^2 \right)\left(\frac{\pi c}{8} \right)$$

(6.12a)

$$= \pi\, m_o\, c^3 / 8$$

The above result indicates that, in the limit of a particle moving infinitesimally close to the velocity of light, its total energy becomes equal to the conventional value ($E = mc^2$), subject to the mass being related to the rest mass as follows:

$$m = m_o \pi c / 8 \tag{6.12b}$$

6.10. Interconnection between Mathematics and Physics in 6-D Spacetime

Table 6.2 summarizes a set of relationships between the dimensionality, geometry, physics, and algebra of systems of interest. As may be naturally expected, there can be seen intimate interconnections between mathematical and physical descriptions of systems in any dimension. It may also be of interest to show that the values of the universal constants ("Anthropic Principle") is the consequence of the 'Minimum Action Principle' under mutual interaction of the basic interactions (strong, weak, EM, gravitation.)

As a corollary, the Anthropic Principle could be simply stated as follows: "The existence of the elementary particles, their basic interactions, and the universal constants, when satisfying certain inherent physical requirements such as the 'Minimum Action Principle,' lead to an environment which is conducive of genesis and support of life, the ultimate evolutionary result of which on the planet earth up till now is the origin of man, capable of observing and analyzing the universe."

Table 6.2. Relation between the Dimensionality, (Euclidean) Geometry, Physics, and Algebra

Dimensionality		Geometry		Physical Process		Algebra
0-D		Point				
1-D	→	Line	→	Action (Linear) Translation Rectilinear (Uniform Velocity) Special Relativity Theory Accelerated Motion Newton's Second Law Einstein's Equivalence Principle General Relativity Theory		
2-D	→	Plane (Euclidean) Geometry	→	Action Rotation	→	Quaternion (Einstein's 1st Mistake: Not using quaternion or hyper-complex algebra, but tensor as space-time curvature)
3-D	→	Solid (Euclidean) Geometry	→	Action Translation υ Rotation		

4-D	→	Atoms/ Molecules	→	Quantum Mechanics Wave Function (Probability in position, time and energy)	
5-D	→	Radioactive Nuclei	→	Quantum Mechanical Tunneling	
6-D	→			Quarks Spin Strong Charge Weak Charge EM Charge Gravitational Charge (Mass): Magnitude of the Quaternion Related to the Particle Mass	Quaternion (Quark-algebra)

CHAPTER 7

Further Discussions of Aspects of Particle- and Systems-Dynamics

"I do not know what may appear in the world, but to myself I seem to have been only like a boy playing on the seashore, and diverting myself in now and then finding a smoother pebble or a prettier shell than ordinary, whilst the great ocean of truth lay all undiscovered before me."

Isaac Newton (1642-1727)

"Twinkle, twinkle, quasi-star
Biggest puzzle from afar
How unlike the other ones
Brighter than a billion suns
Twinkle, twinkle, quasi-star
How I wonder what you are."

George Gamow (1904-1968)

7.1. Introduction

In this Chapter, a unified representation of the four basic interactions is used to develop a unified quantum theory of any type of system, from an elementary particle to cosmological objects, subject to any or all types of interaction(s)—strong, weak, electromagnetic (EM), or gravitational—on the basis of a framework of a six-dimensional spacetime (6-D ST). It is shown how such a unified representation could be utilized in a very simple, natural, and elegant fashion, using the well-familiar quantum mechanical formulism, namely, the Schrodinger equation of quantum mechanics. In a way, this could be treated as an extension of the basic quantum mechanical formulation for interacting systems, not only for elementary particles, and sub-nuclear, nuclear, atomic, and molecular systems, but also for larger systems interacting with gravitational interaction, as well. Thus, this is a trivially simple approach to combine quantum mechanics with the theory of gravitation—or a theory of so-called quantum gravity. But, instead of treating gravitational interaction in isolation, this approach could be applied to any combination of interactions, lending credence to this proposed theory as a **U**nified **T**heory **O**f **E**verything from **P**articles to **I**ntegrated **S**ystems **(UTOEPIS)**.

The concept and algebra, geometry, and matrix formulation of the quaternion plays an important role in this presentation. This, in turn, also establishes the unique physical significance of the quaternion and the associated reality of rotations of axes, planes and coordinate frames in the 6-D ST or, more specifically, in the case of spherically symmetric systems, the space of the spatial radial coordinate (r) and the three time coordinates (t_1, t_2, t_3) or (t_r, t_θ, t_ϕ).

In the present discussion, no reference has been made of spacetime curvature, a concept inherently imbedded in the general relativity; treatment of the space and time continuum in the Euclidean geometry suffices. This, therefore, also leads to a new paradigm concerning the interpretation of mass. Instead of the mass causing the space-time curvature which then being assumed to govern its motion along the geodesic patterns therein, as postulated in Einstein's general relativity theory, it has been shown here how the elementary particle mass could be simply related to the basic interactions themselves, operating in the Euclidean space-time continuum. A self-consistent unification of the classical and quantum physics is thus readily obtained in this representation.

In the discussion below, some common facets of physical laws are examined from a new angle. This includes the inverse square law and Newton's Laws of Motion, and motion in multiple dimensions, for instance.

7.2. The Inverse Square-Law

For an explanation of the inverse square law, we make the physically most simple and natural, and well-familiar, assumption that the source(s) of interaction—the mass (for gravitational interaction) or the charge (for EM interaction)—continuously emit isotropic waves or quantum particles which, when they impinge upon another mass or charge, respectively, cause their respective type of interaction. These are, of course, the gravity and EM-waves or, equivalently, their quanta—gravitons and photons, respectively.

The wave-or particle-manifestations are, of course, equivalent under the well-familiar wave-particle duality principle in quantum mechanics, and observed in numerous physical phenomena. The main point here is that if we assume that the level of interaction between two bodies (masses or charges), say, B and D, are directly proportional to the amount of gravitational waves/gravitons or EM-waves/ photons emitted by B that are intercepted by D (or vice-versa), then this amount would be proportional to the solid angle D makes at B (or vice-versa). Now, if the distance between B and D is r, then the solid angle (Ω) made by D at B is given by

$$\Omega = b/r^2$$

where b is the area of cross-section of D as viewed by B (see Figure 7.1).

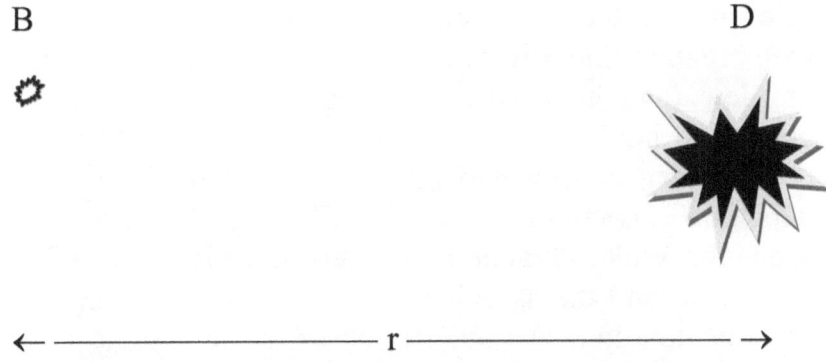

Figure 7.1.   The mechanism of (gravitational or EM) interaction between two bodies B and D.

Here B, as seen by D, is taken as a point-source. For an extended size of B, each point of B can be considered as a point-source of gravity-wave/gravitons or EM-wave/ photons, and the net action obtained as the sum-total of contributions of all the point-sources. Each point-source contributes a solid-angle equal to $d\Omega$, and hence proportional to db/ r^2, db being the contributing infinitesimal portion by the relevant interaction with D. The result is that the total (gravitational or EM) interaction force becomes proportional to a sum of such contributions, each proportional to $1/ r^2$. Obviously, the same conclusion holds for D interacting with B, as is expected from Newton's Third law of equality of action (B acting upon D) and reaction (D acting upon B.)

Clearly, the above view explains the inverse-square law applicable for both gravitational as well as EM interactions, and also provides an answer to the question of "action-at-a-distance." The enormously large speed of the pertinent waves/particles (quanta), c, makes the interaction instantaneous, since the time taken for the wave/quantum particles (r / c) is infinitesimally small for all distances of common experience or relevance. In fact, assuming that quantum uncertainty of positioning and the associated 'vacuum fluctuation' implies that waves are emitted all the time, without interruption, the propagation time, t_r, becomes irrelevant and the interaction can be taken to be present all the time, creating the illusion of 'action-at-a-distance.' Moreover, this view (and hence the resulting inverse-square law) can be assumed to hold not only for gravitational and EM interactions, but for every conceivable interaction, provided the appropriate interaction carriers (say, gluons for strong interaction, $W^{\pm}$ and Z bosons for weak interaction) are considered (in place of gravitons or photons) and the associated special features, if any, are taken into consideration. The unification of the four basic interactions discussed in Chapter 5 explicitly demonstrate the method of obtaining a unified representation of various interaction including, in particular, the inverse-square law of gravitational and EM interactions.

Now we consider different types of motions and related geometrical and mathematical representations, in multiple dimensions, in a simple, unified manner. First, a unified representation of Newton's well-known three laws of motion in terms of 'Action' is presented.

7.3.   One-Dimensional (1-D) Rectilinear Motion

7.3.1. Representation of Newton's Laws of Motion in Terms of 'Action'

In this Section, we demonstrate an underlying unity of the Newton's Laws of Motion. It is interesting to note that all three laws of motion of Newton could be very simply related to the notion of 'Action,' denoted here as A. As a common system parameter, A is defined in elementary physics as

$$A = FLT, \tag{7.1}$$

where F is the force applied to a body or to a system of bodies, L is the distance travelled by the body in question under the influence of this force, and T is the time taken for the motion in question. Using Δl and Δt to denote infinitesimal elements of length and time, respectively, we have the general definitions of velocity, v, and acceleration, f, involved in the above-specified motion as

$$v = \Delta l / \Delta t, \tag{7.2a}$$

and $$f = \Delta v / \Delta t = \Delta^2 l / \Delta t^2, \tag{7.2b}$$

respectively. For a constant velocity, of course, $v = L/T$ (constant), and $f = 0$.

Below, we examine Newton's Laws of Motion in a reverse order to conveniently express them in terms of Action.

Newton's Third Law of Motion

> *"To every action, there is an equal and opposite reaction."*

For two interacting bodies, B and D, let ABD represent the Action of B on D, and ADB, the Action of D on B. The Action for the system as a whole, comprising both the interacting bodies B and D, setting the total Action to be nil, we have

$$A_T = A_{BD} + A_{DB} = 0 \tag{7.3a}$$
$$\text{i.e., } A_{BD} = -A_{DB} \tag{7.3b}$$

The above result is equivalent to the Third Law of Motion, which, therefore, can be generalized for any number of pairs of bodies constituting an entire system. Newton's Third Law of Motion can thus be restated as follows:

> **"For a system as a whole, the total value of 'Action' always remains zero."**

It must be emphasized that this law holds only in absence of an external force (i.e., a force *external* to the system of particles or bodies *as a whole;* that is, any force involved is only the mutual interactions between constituent particles or bodies themselves.)

Newton's Second Law of Motion

"Force is equal to the product of mass and acceleration."

Now, as a special step, we invoke the quantum Uncertainty Principle in order to derive the above (Second) Law of Motion. This may appear strange at the beginning, but it merely goes to show that there is no fundamental difference between classical mechanics and quantum mechanics, except for scale: classical mechanics addresses the physics (particularly, motion) of macroscopic bodies, while quantum mechanics, of microscopic bodies; and, obviously, the definitions of macroscopic and microscopic are themselves subjective and relative, pertaining to a comparison with respect to human experience in day-to-day experience and laboratory experiments for atomic and subatomic scales, respectively, for example. We recall that, according to quantum Uncertainty Principle,

$$\Delta p_x \, \Delta x \sim h \qquad\qquad (7.4a)$$

where Δ denotes the amount of uncertainty in the momentum or in the position variable, $p_x = mv$ is the momentum, and h is the well-known Planck's universal constant, representing a quantum of 'Action.'

It is now convenient for the present purpose to use dimensional analysis. In terms of the dimensions of different quantities involved, we note the dimensions of Action to be a product of the dimensions of force, length, and time (Eqn. 7.1):

$$[A] \rightarrow [F] \times [L] \times [T], \qquad\qquad (7.4b)$$

where the use of *italics* indicates a dimensional relationship (equation), and *[X]* denotes the dimensions of the quantity or physical variable X. Dimensionally, noting that p = mv and v = L/T, Eqn. (7.4a), combined with Eqn. (7.2b) and (7.4b), can be also represented as the following dimensional equation

$$[m] \; x \; [L \; T^{1}] \; x \; [L] \sim [h] = [A] = [F] \; x \; [L] \; x \; [T]$$

i.e., $[F] = [m] \; x \; [LT^{2}] = [m] \; x \; [f]$

Choosing a suitable system of units according to which a unit force (F = 1 dyne, say) is the one which produces a unit acceleration (f=1 cm/sec^2) in a body of unit mass (m=1gm), as the conventional definition of force in the cgs system indeed is, the above dimensional equation can be put in the form of a mathematical equation:

$$F = m \; f, \tag{7.4c}$$

which is Newton's Second Law of Motion.

Rewriting Eqn. (7.4c) in the form

$$(F - mf) \; x \; L \; T = 0.$$

the (Second) Law of Motion is also equivalent to the statement:

"For a body subject to an external force, the resulting motion is such that the net Action involved is zero."

Newton's First Law of Motion

"Unless and until acted by an external force, a body remains in the state of rest if initially at rest, or continues to move with a constant velocity."

In Eqn. (7.1), setting the (external) force to be zero, and using Eqn. (7.4.c),

$$F = mf = 0, \tag{7.5a}$$

we must have, for m ≠ 0,

$$f = \Delta v / \Delta t = 0, \tag{7.5b}$$

and also, from Eqn. (7.1),

$$A = 0, \tag{7.5c}$$

Eqn. (7.5b) necessarily implies that $v = 0$, or $v = $ constant, depending on whether the body is initially at rest or moving at a velocity v (constant in magnitude as well as in direction.) This is the First Law of Motion. We have thus effectively proven that in absence of an external force, when a body remains at rest or continues moving with a constant velocity (as stated in Newton's First Law), the Action *pertaining to the body in question* is zero.

7.3.2. A Unified Representation of Newton's Laws of Motion

The above set of results can be concisely summarized in a fashion that could be taken as a unified representation of Newton's all three laws of motion in the form a single law expressed in terms of the Action variable as defined by Eqn. (7.1):

> *"For a particle (or body) or to a system of particles (or bodies), the net Action pertaining thereto always vanishes."*

i.e., $A = 0$ (7.6)

Note that the above "Law of Action" explicitly embodies Newton's First Law of Motion (Eqn. 7.4.c) and the Third Law of Motion (Eqn.7.3a); while *the very definition of Action itself directly leads to the derivation of the Second Law of Motion, Eqn. (7.4c).* Furthermore, we have also been automatically led to a synergetic treatment of matter of all (microscopic to macroscopic) size and scale, or a unification of classical mechanics and quantum mechanics from the above discussion, an added bonus by virtue of treatment in terms of Action.

The above essentially pertains to rectilinear motion, and further generalization and interconnection may be expected to describe rotational (2-dimensional) motion and other physical phenomena in terms of the Action variable. In passing, one may recall Fermat's Principle in optics stating that the rectilinear propagation of a light ray (wave) with a constant velocity (as in a vacuum, i.e., without any external force acting on it that may cause reflection, refraction, interference, etc.), c, is tantamount to an optimization under which the applicable 'Action' variable vanishes. Mathematical generalization of the above type of *Vanishing Action Principle* to possibly include a wide range of other physical phenomena and theories (e.g., special theory of relativity, general theory of relativity, Lorentz transformation formula, gravitational bending of light ray, etc.) could also be explored, expectedly with much dividends.

7.4.   Two-Dimensional (2-D) Motion

7.4.1. General

Here, we first consider rotational motion which is typically a two-dimensional (2-D) motion.

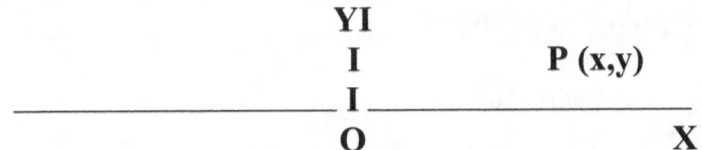

Figure 7.2.   Rotational Motion

Now, as shown in Figure 7.2, consider a rotation of the linear axis OX by an angle θ (taken positive for anti-clockwise rotation), about any axis (OZ, not shown in Fig. 7.2), defining a plane (x,y) perpendicular to OZ, such that the three axes OX, OY, and OZ, form a (right-handed) Cartesian coordinate system. A circular motion of a particle P of mass m around the circular path with radius $OP = r$ and uniform angular speed $w = \frac{d\theta}{dt}$, will now be considered in terms of Action (A).

For an infinitesimal time dt, the motion of the particle could be assumed to be approximately linear with (infinitesimally constant) velocity v,

$$v = \Delta s / \Delta t = r \, \Delta w / \Delta t$$

and acceleration (in the linear direction) f,

$$f = \Delta v / \Delta t$$

Of course, this type of (rotational) motion arises in the presence of a centripetal force (balanced by an equal and opposite centrifugal force)

$$F_c = \frac{mv^2}{r} = mrw^2$$

So that the (differential) Action, assuming an infinitesimal time dt to be required for traversing a unit distance ($ds = 1$), is

$$dA = F_c dt = mrw^2 dt \; (\neq 0) \tag{7.7a}$$

The angular momentum is

$$p_w = mvr = mr^2 w \tag{7.7b}$$

For a microscopic system (e.g., elementary particle), assuming this angular momentum (not to be confused with spin) to be an integral multiple of $\frac{h}{2}$, where h is the Planck constant, we can write

$$p_w = mr^2 w = n\frac{h}{2}, \quad n = 0, 1, 2, 3, \ldots \tag{7.7c}$$

Now, from Equations(7.7a) and (7.7c),

$$dA = \left(mr^2w\right)\frac{w}{r}dt = \frac{nh}{2}\frac{d\theta}{r}$$

Hence, the Action per unit angle is given by

$$A_\theta = \frac{dA}{d\theta} = \frac{nh}{2}\frac{1}{r} \qquad (7.8a)$$

Thus the Action corresponding to a full revolution (2π radians) is

$$A_0 = 2\pi A_\theta = nh\frac{\pi}{r} \quad ; \qquad (7.8b)$$

while the Action per unit length of arc, A_s, and the Action per unit area enclosed by this particle orbit, A_a, are

$$A_s = \frac{A_0}{2\pi r} = n\left(\frac{h}{2}\right)\frac{1}{r^2} \qquad (7.8c)$$

and

$$A_a = \frac{A_0}{\pi r^2} = \frac{nh}{r^3} \qquad , \qquad (7.8d)$$

respectively.

7.4.2. Electromagnetic (EM) Interaction

In particular, in the case of the Bohr model of the atom, the central force is the electromagnetic (EM) attraction between the orbiting electron of charge -e, and the atomic nucleus with protons, and hence with charge +Ze, where Z is the atomic number. The centripetal force exerted on the orbiting electron (neglecting the EM forces due to other orbiting electrons) is

$$F_e = (-) Ze^2/r^2 = m\, r\, w^2 \tag{7.9a}$$

since the centripetal force must equal the centrifugal force for a dynamic equilibrium of the orbiting electron.

Combining Eqns. (7.7c) and (7.9a), we have

$$\frac{Ze^2}{r^2} \cdot \frac{r}{w} = n\frac{h}{2} \tag{7.10a}$$

Hence, we are led to a set of quantized orbital radii for the electron, viz,

$$r_n = \frac{2Ze^2}{wh} \frac{1}{n} \tag{7.10b}$$

Substituting for w (Eqn. 7.9a), viz.,

$$\frac{1}{w} = \left(\frac{mr^3}{Ze^2} \right)^{\frac{1}{2}} \quad ,$$

we have

$$r_n = \frac{2Ze^2}{hn} \left(\frac{mr^3}{Ze^2} \right)^{\frac{1}{2}} = \frac{2\sqrt{Ze^2 m}}{hn} r_n^{\frac{3}{2}} \quad ,$$

i.e.,
$$r_n = \left(\frac{hn}{2\sqrt{Ze^2 m}} \right)^2 = \left(\frac{h^2}{4Ze^2 m} \right) n^2 \tag{7.10c}$$

The corresponding quantum (EM) potential energy levels can, therefore, be simply written as

$$E_n = -\frac{Ze^2}{r_n} = -Ze^2 \left(\frac{4Ze^2 m}{h^2} \right) \frac{1}{n^2}$$

i.e.,
$$E_n = -\frac{4Z^2 e^4 m}{h^2} \cdot \frac{1}{n^2} \tag{7.10d}$$

A change (drop) of the electron from the nth to the pth (quantum) energy level, $p < n$, then corresponds to the emission of radiation of frequency v_{np} given by

$$v_{np} = \frac{E_n - E_p}{h} = \frac{4Z^2 e^4 m}{h^3} \left(\frac{1}{p^2} - \frac{1}{n^2} \right)$$

(7.10e)

This result for atomic radiation frequency spectrum was first derived empirically and subsequently from the quantum mechanical theories, namely, Schrodinger's wave mechanics yielding differential equation and Heisenberg's matrix mechanics. Here we have obtained the same result without any recourse to any partial differential or matrix equation, but only by using the principle of the Action (here, the angular momentum) being an integral multiple of h/2.

7.4.3. Gravitational Interaction

Due to the commonality of the inverse-square law between the EM and gravitational forces, and noting that the gravitational (attractive) force on a body of mass m due to another body of mass M is equal to

$$F_g = (-)GMm / r^2 ,$$

(7.11a)

the above results are, in principle, also applicable to gravitational interaction potential

$$V_g = GMm / r ,$$

(7.11b)

if we replace the factor (Ze^2) by the factor (GMm) in the above equation (Eqn. 7.10). Thus we are then led to the expected frequency of the **gravitational waves** emitted by the revolving body.

Again, it should be emphasized that *a convergence and unification of classical and quantum mechanics, a theory of quantum gravity (gravity waves or gravitons)* is embodied in the above presentation.

Under what circumstances could the law of universal gravitational attraction become one of repulsion? What if, under certain mechanism, the mass were to be represented as an imaginary quantity! Substituting (im) for mass (m), $i = \sqrt{(-1)}$, in the formula, we have

$$F_g = (-)G\ (iM)\ (im)\ /r^2 = G\ M\ m\ /r^2 \quad , \qquad (7.11c)$$

which represents a repulsive force (since M and m are both positive.) The imaginary (instead of the usual real) part representation of mass could be said to be the result of certain rotation in the 6-D ST, as in the definition of the quaternion. The problem of the dark energy responsible for making the cosmological expansion speed up may be 'explained' in this type of transformation, since the 'repulsive gravitational force' would make the galaxies and galaxy-clusters to fly apart with increasing speed.

Another interesting representation of the gravitational force results if we write $(M\ m\ /\ r^2)$ as the product $(M/r) \times (m/r)$, showing that the force is proportional to the product of the 'masses per unit distance' of separation between them.

7.5. Further Considerations: 2-D Case

We now consider the case of 2-D motion, thereby also highlighting the interrelations among the physical significance of the unique mathematical quantities π, i and e in the context of introduction of extra dimensions. The related descriptions in terms of vector analysis, complex and hyper-complex (quaternion) representation, rotation, and matrix algebras are also presented.

7.5.1. Action per Radian (ħ)

If we assume the Planck constant as representing one unit of 'Action' corresponding to a full circular (2π radians) rotation or spin, the parameter

$$\hbar = \frac{h}{2\pi}$$

(7.12a)

is to be regarded as the 'Action' per unit radian.

7.5.2. Pi (π) As an Operator (Generating 2-D S from a 1-D space)

Also, rotation of a radial line of length r by a full circular (2π radians) generates a perimeter $s = 2\pi r$ spanning a 2-D coordinate space (x, y) convertible to a 2-D polar coordinate (r, θ). Hence, as a fundamental attribute, π can be regarded as the operator for introduction of the second dimension (1-D $\rightarrow$ 2-D) and the radius, r, is equal to the arc-length per radian:

$$r = \frac{s}{2\pi}$$

A 2-dimensional (2-D) space can be described in a number of alternative ways, as illustrated below. In particular, this applies not only to ordinary space (2-D S), such as the (x,y)-plane, but also to the 2-D time-space, e.g., (t_x, t_y). The following discussion is focused to the (x,y) plane only, however, for brevity. While these results are elementary, their relevance in the context of time-space and further generalizations to higher dimensions in connection with rotation and quaternion should be kept in mind.

7.5.3. 2-D Vector Space

A 2-D vector can be represented in terms of the unit vectors $\underline{x}$ and $\underline{y}$ along the x- and y-axes respectively, so that $\vec{r} = x\,\underline{x} + y\,\underline{y}$: with $r^2 = x^2 + y^2$ (Pythagoras Theorem);

i.e., $r = \left|\vec{r}\right| = \sqrt{x^2 + y^2}$

and $\theta = \tan^1 = \dfrac{y}{x}$

is the angle $\vec{r}$ makes with the x-axis.

7.5.4. Complex Plane

The (x,y) plane could be regarded as a complex plane, with the complex variable z

$$z = x + iy, \ i = \sqrt{(-1)}$$

so that

$$z^2 = x^2 + y^2$$

and $|z| = (x^2 + y^2)^{1/2}$

7.5.5. Quaternion and Pauli Matrices

In the present (2-D) case, setting (see Section 2.8) $\alpha_0 = \alpha_3 = 0$ and $\alpha_1 = x$, $\alpha_2 = y$, so that

$$\tilde{Q}(2) = xi + yj$$

And $\left|\tilde{Q}(2)\right| = \sqrt{x^2 + y^2} = r$

It should be noted that a general quaternion (Eqn. 2.4e) corresponds to a 2×2 unitary matrix U

$$U = \begin{bmatrix} \alpha_3 & \alpha_1 - i\alpha_2 \\ \alpha_1 + 1\alpha_2 & -\alpha_3 \end{bmatrix} = \alpha_1 P_1 + \alpha_2 P_2 + \alpha_3 P_3 \qquad , \qquad (7.13)$$

where P_1, P_2, P_3 are the 2×2 Pauli matrices

$$P_1 = \begin{bmatrix} 0 & 1 \\ -1 & 0 \end{bmatrix}, \cdot P_2 = \begin{bmatrix} i & 0 \\ 0 & -i \end{bmatrix}, \cdot P_3 = \begin{bmatrix} 1 & 0 \\ 0 & -1 \end{bmatrix} \qquad (7.14)$$

which correspond to the unit vectors $\hat{i}, \hat{j}, \hat{k}$ along the axes (x, y, z) and which, together with the 2×2 unit matrix

$$I = \begin{bmatrix} 1 & 0 \\ 0 & 1 \end{bmatrix} \qquad ,$$

themselves satisfy the properties of a quaternion:

$$P_1^2 = P_2^2 = P_3^2 = -I = P_1 P_2 P_3 \qquad (7.15a)$$

$$P_1 P_2 = -P_2 P_1 = P_3 \qquad (7.15b)$$

$$P_2 P_3 = -P_3 P_2 = P_1 \qquad (7.15c)$$

$$P_3 P_1 = -P_1 P_3 = P_2 \qquad (7.15d)$$

Further,

$$UU = \begin{bmatrix} \alpha_3 & \alpha_1 - i\alpha_2 \\ \alpha_1 + i\alpha_2 & -\alpha_3 \end{bmatrix} \begin{bmatrix} \alpha_3 & \alpha_1 - i\alpha_2 \\ \alpha_1 + i\alpha_2 & -\alpha_3 \end{bmatrix} = \begin{bmatrix} \alpha_1^2 \alpha_2^2 \alpha_2^3 & 0 \\ 0 & \alpha_1^2 \alpha_2^2 \alpha_2^3 \end{bmatrix} = \left(\alpha_1^2 \alpha_2^2 \alpha_2^3 \right) I$$

$$(7.16a)$$

In the present 2-D case, putting $\alpha_0 = \alpha_3 = 0$, $\alpha_1 = x$, $\alpha_2 = y$

$$U(2) = \begin{bmatrix} 0 & x-iy \\ x+iy & 0 \end{bmatrix} = xp_1 + yp_2$$

$$= x \begin{bmatrix} 0 & 1 \\ -1 & 0 \end{bmatrix} + y \begin{bmatrix} i & 0 \\ 0 & -i \end{bmatrix}$$

$$= \begin{bmatrix} 0 & x \\ x & 0 \end{bmatrix} + \begin{bmatrix} iy & 0 \\ 0 & -iy \end{bmatrix} = \begin{bmatrix} iy & x \\ -x & -iy \end{bmatrix} \tag{7.16b}$$

7.5.6. 2-D Rotation

In the 2-D case, the anti-clockwise rotation of the coordinate axis x by an angle θ, about the z-axis, transforms the values (x, y) into (x', y')

$$\begin{bmatrix} x' \\ y' \end{bmatrix} = \begin{bmatrix} \cos\theta & \sin\theta \\ -\sin\theta & \cos\theta \end{bmatrix} \begin{bmatrix} x \\ y \end{bmatrix} \tag{7.17}$$

7.5.7. Polar Coordinate Representation: The Significance of i and e.

If (x,y) corresponds to the polar coordinates (r, φ):

$$x = r\cos\varphi, \, y = r\sin\varphi, \, r = \sqrt{x^2 + y^2}, \, \varphi = \tan^{-1}\frac{y}{x}$$

then

$$z = x + iy = re^{i\varphi} = r(\cos\varphi + i\sin\varphi)$$

Clearly, $r\cos\varphi$ and $r\sin\varphi$ represent two simple (sinusoidal) waves along orthogonal directions (x- and y-axes, respectively) which are out of phase by $\frac{\pi}{2}$. This is reminiscent of the plane *EM* wave traveling (with the velocity c) in the *Z*-direction, with the x- and y-axes carrying the electric and magnetic fields, respectively.

7.6. Three-Dimensional (3-D) S Case

Generalization of the above formalism to the case of 3-D rotation is obviously possible. For brevity, here we omit the discussion of a 3-D S and 6-D ST rotation and other modes of motion.

A general comment on the nature and relevance of the orthogonal time coordinates must be made to avoid any confusion. At the first sight, the notion of 3-D T concept may appear vague or mysterious. However, the orthogonal time coordinates simply refer to the time measure involved in the orthogonal motion, with the line of sight or the direction of observation being taken as the main time coordinate used for common reference. Often, most of the discussions in connection with cosmological motion are confined to the direct radial motion (such as the Hubble expansion.) However, the cosmological bodies may also be endowed be motion in a plane perpendicular to this radial direction. The characterization (velocity, acceleration, rotation, etc.) of motion in this orthogonal plane, for an arbitrary case, must be made with respect to time variables (t_θ and t_ϕ, say) which, together with the usual (radial) time coordinate, t_r, form three independent degrees of freedom, which is tantamount to a three-dimensional time T).

By the same token, energy polarized in the transverse direction, as mentioned in connection with the dark energy, is simply waves polarized in the transverse direction. For a plane wave, the variations in the electric and magnetic fields take place in this transverse plane. For a general case, similarly, a three-fold degrees of freedom, or a three-dimensional time-based description of the field(s) need to be employed.

As is well-known, a quaternion represents a rotation in a four-dimensional space (say, in the 4-D ST of r, t_r, t_θ, t_ϕ) and its representation in terms of the Pauli matrices, generalized to the Dirac matrices in a relativistic case, provides a connection with spin and the relativistic theory of electron, for example. Generalization for other elementary particles by using the corresponding quaternion is obviously possible in this framework.

CHAPTER 8

Origin of Spin

"The population of scientific doctrines is producing as great an alternation in the mental state of society as the material application of science are effecting in its outward life. Such indeed is the respect paid to science, that the most absurd opinions may become current, provided they are expressed in language, the sound of which recalls some well-known scientific phrase."

James Clark Maxwell (1831-1879)

8.1. Introduction

It is well-known that as the plasma cooled enough after the big bang, electrons were captured by protons to form Hydrogen atoms under the electromagnetic (EM) interaction, with the proton at the center forming the nucleus while the electron assumed spin as well as orbital revolution. Similarly, in the case of the solar system, planets were formed which became trapped in the gravitational field of the sun and began revolving around the sun, together with spin motion. Here we develop a mathematical expression for the expected spin motion for the electron in a hydrogen atom, and also for the earth around the sun, and numerically estimate the value of the spin. For this purpose, the principle of conservation of angular momentum is invoked. It is assumed that a freely moving object (electron or a planetary body) moving with slow enough speed in space and passing close enough with respect to a central force-field (EM or gravitational) is captured by the central body to start revolving around it, in combination with a spin angular momentum.

8.2. Mathematical Model

Let us assume that a body A of mass m moving freely in space with a uniform velocity V passes close to a central body B, with a distance of closest approach equal to R, and is captured by B such that A starts revolving around B in a circular orbit of radius r and with a spin angular motion corresponding to nh/2, where h is Planck constant, and n is an integer. This last assumption is based on the quantum theory of spin angular momentum as in the case of the Bohr atomic model.

Figure 8.1 illustrates the system of bodies A and B.

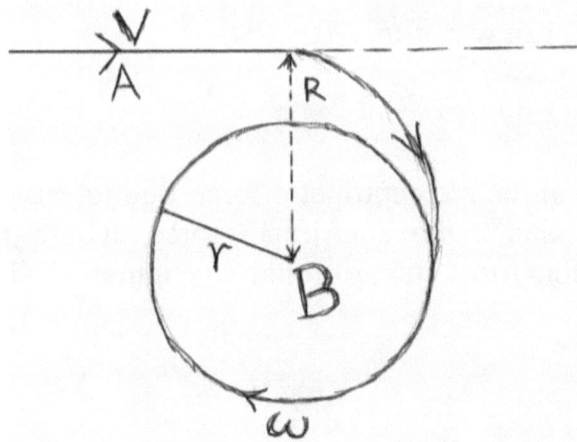

Figure 8.1.   The geometry of a central force-field due to the body or particle B capturing a free-streaming body or particle A in an orbital motion, in combination with a spin angular momentum.

The initial angular momentum of A with respect to B, before its capture to form a bound system, is, of course

$$\Delta_0 = m\,V\,R \qquad\qquad (8.1a)$$

After the capture the final angular momentum of A with respect to B is composed of two components, corresponding to the orbital (circular) motion of revolution of A around B, and the spin motion of A, respectively. The first component is equal to $(m.v)\,r = mwr^2$, where v is the linear velocity of A in its orbital revolution, and $w = v/r$ is the angular velocity of A around B; while the second component is simply $nh/2$, as assumed above. Thus the final total angular momentum is given by the sum

$$\Delta_f = m\,w\,r^2 + nh/2 \qquad\qquad (8.1b)$$

Equating the initial angular momentum to the final one (Conservation of Angular Momentum), we get

$$n\,h\,/\,2 + m\,w\,r^2 = m\,V\,R$$

Hence, $n = (2m\,/\,h) \times (V\,R - w\,r^2)$ \qquad\qquad (8.1c)

Now, the attractive centripetal force due to the central force field must balance the centrifugal force due to the uniform circular motion for a stable orbital configuration. This leads to the equation:

$$F_0\,/\,r^2 = m\,v^2\,/\,r \qquad\qquad (8.2a)$$

where $F_0 = [\ (-)\ e^2$, for the hydrogen atom (EM interaction)], and

$F_0 = [(-)\ GMm$ interaction, for the earth (gravitational interaction)]

$$(8.2b)$$

Here, (+ e) is the electric charge of the proton at the center, (- e) is the electric charge of the electron with mass m; while in the planetary case, M is the mass of the sun, and m is the mass of the earth. Combining Eqn. (8.1c) and Eqn. (8.2a), we obtain the desired expression for the quantum number n:

$$n = (2m/h) \cdot [VR - (F_0 \, r \, / \, m)^{1/2}] \qquad (8.3)$$

We can now evaluate n for the EM (hydrogen atom) and gravitational (planet earth around the sun) cases. A rough computation indicates that under proper conditions, n = 1 for the hydrogen atom; while, of course, n takes a very large value in the planetary case, as expected.

8.3   Numerical Calculations

For illustrative approximate numerical calculations, we use the values of the universal constants, together with appropriately assumed values of other parameters, as indicated below:

8.3.1. Electromagnetic Interaction: The Hydrogen Atom

The values of the constants are:

(+/-) e = proton (+) electric charge = electron (-) electric charge = 4.8×10^{-10} esu
m = electron mass = 9.11×10^{-28} gm
h = Planck constant = 6.63×10^{-27} erg.sec

We also arbitrarily assume

r = the Bohr radius for hydrogen atom = (1.058×10^{-8}) cm (and R = 0.5 x r)
V = 1.012×10^{10} cm/sec
Substituting these numerical values in Eqns. (7.1c), (8.2a), and (8.2b) for the EM case, we obtain

$$n = (2 \times 9.11 \times 10^{-28} / 6.63 \times 10^{-27}) [1.012 \times 10^9 \times 5.29 \times 10^{-9} -$$

$$4.8 \times 10^{-10} (10.58 \times 10^{-9} / 9.11 \times 10^{-28})^{1/2}]$$

i.e., $n = 1$

In other words, the spin of the electron is $(1 \times h/2) = h/2$, as expected.

Of course, some adjustment of the numerical values of V (velocity of the free electron) and R (the distance of closest approach with respect to the hydrogen nucleus, or the proton) has been incorporated in order to obtain n = 1; but the above calculation generally validates the basic model for generation of spin, and also yields some insight as to the physics (typical velocity of the free-streaming electron and the distance of the closest approach in order for the electron to be captured by the proton to form the bound system (hydrogen atom)) involved. For instance, in the above example, it indicates that for a free electron moving at a speed equal to one-tenth of the velocity of light, the estimated distance of minimum approach (capture distance) is less than the radius of the hydrogen atom, but roughly 100 times the radius of the nucleus. Obviously, other combinations of the electron velocity and the distance of minimum approach are also possible for the capture to take place, subject to the condition VR = 5.3549 ; hence V is inversely proportional to R.

8.3.2. Planetary System: The Case of the Sun and the Earth

Considering the revolution of the earth around the sun, we have the following constants and other variables:

G $= 6.67 \times 10^{-8}$ cm^3/gm.sec
m   = mass of the earth $= 5.98 \times 10^{27}$gm
M   = mass of the sun $= 1.99 \times 10^{33}$ gm
r   = earth's radius of revolution around the sun (assuming circular radius)

        $= 1.5 \times 10^{13}$ cm

h   = Planck's constant = 6.63×10^{-27} earg.sec

Also we arbitrarily assume

R   = the distance of closest approach = orbital radius r = 1.5×10^{13} cm

V $= 2 \pi r / T = 2 \times 3.14 \times 1.5 \times 10^{13} / (365 \times 24 \times 60 \times 60)$ cm/sec

        $= 3.15 \times 10^{6}$ cm/sec,

where T above is the orbital period of the earth around the sun (= 1 year.)

Substituting the numerical values and simplifying, we obtain

    $n = 1.35 \times 10^{74}$

and $n h / 2 = 8.95 \times 10^{47}$ earg.sec

then yield the spin angular momentum of the earth for its rotation about its axis. Such a high value of the spin angular momentum is of course expected due to the large mass of the earth.

Appendix A

On the Matter of Time

A.1. Historical Background

As is well known, the exact nature of space and time—especially of time—has been contemplated from the earliest days of development of human civilization, philosophy, and particularly, science. Physicists and mathematicians including Euclid (325-265 BCE), Descartes, Galileo, Newton, Leibniz, Lagrange, and, of course, Einstein, have pondered upon the mystery of space and time1, starting from the fifteenth century up till now. Einstein was the first to introduce the concept of time as an "imaginary" variable, thereby introducing the concept of a four-dimensional space-time (4-D SP) continuum, in terms of the four mutually orthogonal coordinates $(x, y, z, ict)^{*}$.

However, as has been discussed before in this book, Einstein's choice of aligning the direction of the relative velocity along a single coordinate-axis, say, the x-axis, as is conventional in special relativity, effectively reduces space-time to be only two-dimensional (x, ict_x) or (r, t_r). Moreover, special relativity

---

* For a comprehensive discussion of various viewpoints of space and time historically [see Judith V. Grabiner, "Why Did Lagrange 'Prove' the 'Parallel Postulate,'" the American Mathematical Monthly, Vol. 116, No. 1, January 2009, pp. 3-18].

does not particularly dwell on the physical significance of the crucial quantity *ict*, except for focusing on the invariance of the 4-D (or, effectively, 2-D) interval $x^2 + y^2 + z^2 + (ict)^2$, subject to the assumption of a universally constant velocity of light, c.

By a general treatment of the relative velocity vector, the concept of a three-dimensional time vector (3-D T), and hence of a six-dimensional spacetime continuum (6-D ST), has been used in this book. This permits a logically and physically meaningful interpretation of the interactions to correspond to rotation in the 6-D ST, and also lends to a simple model for the elementary particle masses as a linear function of the sum of the squares of the interaction constants. The simple formulism involved also leads to prediction of the masses of the Three Higgs bosons, one of which has been apparently detected by the Large Hadron Collider (LHC) at CERN in Geneva, Switzerland, as announced on July 4, 2012.

A.2. Scriptural Connnection

The measure of Time in the context of the origin and fate of the universe, including the Big Bang theory, is an intriguing element of physics and cosmology. In this connection, it is worthwhile to note that every religious tradition of the world propounds a certain picture or model of the origin of the universe, and associated description of time and time-scale. The Christian theological picture is well familiar in the Western world: *("God made the heaven and the Earth . . . in 7 days . . ."*, Genesis 1:1.) Perhaps not so well known is the Hindu neo-scientific-cum-mythological picture and description comprising a cyclical creation and dissolution of the universe by the Divine Creator (*Brahmaa*—not to be confused with '*Brahman*', the Single, Universal, Ultimate Reality or the Supreme Being—God. *Brahmaa* is one of the Hindu Trinity, while '*Brahman*' refers to the Hindu concept of monotheism propounded in the Upanishads—forming part of the vast Hindu scriptures.) Colossal cosmic scales of Time are involved in this picture with units called "Brahmaa's Day and

Night" and its sub-multiples and super-multiples, spanning billions of years, which, astonishingly, compare well with the modern estimates of the age of the universe according to the Big Bang theory, and beyond. For curious readers, the Hindu cosmic scales of Time are briefly summarized below.

Time in the Hindu Scripture:

According to the Hindu Puranic scripture **Shrimad Bhagavatam**, Time has no beginning, and no end; yet the universe undergoes periodic rebirth and total dissolution (*'Pralaya'*), as Lord Brahmaa becomes awake during the Brahmaa-day and goes to sleep during the Brahmaa-night. How is the Lord Brahmaa's Time related to the human ("earth") time? In terms of the cosmic time-unit, symbolically represented here as 'U', this relationship is expressed, in terms of a basic unit here simply labeled as *1 U = 432,000 (Earth or human) years*, as spanning four *Yugas* (cosmic periods) called the **Sat-Yuga, Treta-Yuga, Dwapar-Yuga, and Kali-Yuga,** respectively.

The lengths of the four Yugas are delineated as follows:

Kali-Yuga   = 1 U = 432,000 (human) years
Dwaapar-Yuga = 2 U = 864,000 (human) years
Tretaa-Yuga   = 3 U = 1,296,000 (human) years
Sata-Yuga   = 4 U = 1,728,000 (human) years

The sum of the periods of the four Yugas is called a Chaturyugee or Maha-Yuga. Hence,

1 'Chaturyugee' = 10 U = 4,320,000 (human) years

= 4.32 million (human) years

Now, 1000 'Chaturyugees' make up One (1) Day of Brahmaa, and an equal length of Time constitutes One (1) Night of

Brahmaa. A set of 360 such Brahmaa Day-plus-Night periods define One (1) Brahma-Year, called a 'Kalpa'; while 100 Kalpas constitute One (1) 'Maha-Kalpa' of cosmic Time. (Note: the number 360 is likely associated with the complete span of longitudinal type variarible), i.e.,

$$
\begin{aligned}
\text{One (1) Day} &= \text{One (1) Night} = 4.32 \times 1000 \text{ million (human)} \\
\text{of Brahmaa} &\quad \text{of Brahmaa} \qquad \text{years} \\
&\qquad\qquad\qquad\qquad = 4.32 \text{ billion (human) years} \\
1 \text{ Kalpa} &= (4.32 + 4.32) \times 360 \text{ billion (human) years} \\
&= 3110.4 \text{ billion (human) years} \\
&= 3.1104 \text{ trillion (human) years}
\end{aligned}
$$

$$
\begin{aligned}
1 \text{ Maha-Kalpa} &= 100 \text{ Kalpa} \\
&= 311.04 \text{ trillion (human) years (Life-span of } \textit{Brahmaa.)}
\end{aligned}
$$

The above pattern of cosmic Time-Cycles (of Maha-Kalpa) continues indefinitely, without beginning, and without end.

A.3.   Hindu Cosmic Time-Scale: Manus and Vyāsas

A synopsis of the cosmic time-scale as depicted in the Puranas has been presented above. In a Brahmā's day there are fourteen **Manus** (sages appointed by Brahmaa to oversee the Creation and evolution of Man.) The duration of each Manu's period is called a ***Manvantara.*** In each *manvantara,* a new *Indra* rules over the *Svargaloka*. The name of the present Manu is *Shraddhadeva Vaivasvata*, and the name of the present Indra is *Purendara*, with wife *Shachi* (Vishnu-Purāna 3.2.50; 3.1.30-31.) In each Dwāpara-Yuga, there appears a **Vyāsa** who organizes the ONE VEDA into four volumes to facilitate their study. The 28 Vyāsas who appeared during the 7th manvantara are also enumerated in the scripture (Vishnu-Purāna 3.1.35-42.)

In each Manvantara, there are seventy-two cycles of the Mahāyuga. In the **present manvantara**, which is seventh in succession, twenty-seven Mahāyugas have already elapsed and the present cycle corresponds to the 28th Mahāyuga.

i.e.,   1 Manvantara = 1 Day of   = 4.32 x 1000
                        Brahmaa / 14   / 14 million
                                           (human) years
                                         = 308.5714286
                                           million
                                           (human) years

Alternatively, 1 Manvantara = 72 Maha-Yugas = 72 x 4.32
                                               million
                                               (human) years
                                             = 311.04 million
                                               (human) years

The discrepancy in the two numerical values of the time-period of one Manvantara is due to certain approximations.

A.4.   Hindu Cosmic Time-Scale and Modern Physics (the Big Bang Theory, Multiverse)

It may be noted that, according to the well-known **Big Bang theory of modern physics**, the age of the solar system (and of the earth) is estimated to be 4.3 billion years. This, according to the Hindu scriptures (Puranas), is precisely equal to the length of One Day of Brahmaa, when (the latest cycle of) Creation took place and is continuing. However, the Hindu cosmic time-scale referring to an infinite recurrence of such periodic time-scale in larger time-units such as Kalpa and Maha-Kalpa is far more vaster than the perception of time in modern physics, which limits itself to only the latest cycle.

A distinction must be made between the **cosmic Yugas** consisting of astronomical lengths of time, and the recent

historical (dynastical) Yugas, for example, the *Treta-Yuga of Rama*, and the *Dwapar-Yuga of Krishna;* since the use of the same terminology of the two types of time-scales or periods (cosmic and historical) may be confusing. Similarly, the question of Incarnation also must be viewed with proper time-scales (cosmic versus historic), and a parallel could be seen with respect to the newly advanced **Darwin's theory of evolution,** as there appeared, according to the Hindu scriptures, Turtoise-Form (*Kachhapavataar*), Fish-Form (*Matsyavataar*), Half-human-Form (*Narsinhavataar*), Humanoid-Form (*Vamanavataar*), and then a set of Human-Form (*Parasuram, Ram, Krishna, Buddha*) Incarnations, in that sequence. Such a sequence clearly reflects the evolutionary pattern on the earth now widely accepted.

According to modern physics and cosmology, the approximate age of the known or observable universe, comprising about 200 billion galaxies, each, in turn, comprising about 200 billion stars like our sun, is 13.5 billion (human) years. That is, the universe, according to the latest theories of physics, was "born" from a microscopic nucleus of matter about 13.5 billion (human) years ago. Before the Big Bang, Time and Space are said to have no meaning; Time (or Space) simply did not exist prior to the Big Bang.

Who brought about the cataclysmic event, or what existed before the Big Bang are questions outside the domain of physics. The question of existence or of the source of Creation (of the universe)—the question of 'God'—are of no concern to science as it is practiced today (though such questions are beginning to appear at the horizon of physics; many physicists are getting increasingly interested in the questions of what was before the Big Bang; and how was the universe born; or how it might end.)

That single point, which scientists call singularity, where all the mass and energy of the entire universe was stored, is called in the Vedic language *'Hiranyagarbha'*—literally meaning the Womb of Energy. 'In the Beginning was *Hiranyagarbha*, the Seed of elemental Existence, in the most condensed form.' The universe itself is called 'Brahmaand' which literally means the Egg generated by Brahmaa.

A.5. The Current Phase of the Universe

The Puraanic Concept of Creation and Dissolution has been briefly described above. As mentioned, the largest unit of the cosmic Time-dcale, equal to one Brahmaa-Life, is called a *Maha Kalpa*. After one *Maha Kalpa*, there is a complete Dissolution of the *Brahmaand* (the primordial Seed of the manifested Universe) itself. This grand cosmic episode is called *Prakrati Pralaya of the Brahmaand* [Bhaagavatam (12/4/2-6)].

According to the Hindu cosmic calendar, Lord Brahmaa has completed 50 (Brahma-) Years of His life and is in the 1st Day of the 51st Year of the current *Maha Kalpa*. Therefore, Brahmaa's Age, or the elapsed time in the *Maha Kalpa* in terms of man's time scale (human years) is

= 50 x 720 (Days and Nights in one Year of Brahmaa) x 4.32 billion years (One Day of Brahmaa) + Time elapsed in the 51st Year

= 155.52 trillion (human) years + time elapsed in the 51st year.

For the calculation of the elapsed time in 51st year, let us consider the following:

As far as the present universe (the time elapsed in the 51st Year of Brahmaa's Life) is concerned, it has already completed six out of fourteen M*anvantara*, Twenty-seven out of Seventy-two *Chaturyugees* in the seventh *Manvantara*, three *Yugas* in the Twenty-eighth *Chaturyugi*, and by the (human-) year 2020 would be completing 5114 (human-) years in the Fourth *Yuga*, known as *Kali-Yuga of the present cosmic cycle*.

6 M*anvantara* = 6 x 308.5714286 million (human) years = 1,851.4285716 million (human) years

27 *Charuryugee* = 27 x 4.32 million (human) years = 116.64
   million (human) years

3 *Yugas: Satyuga, Treta and Dwapar* = (4 + 3 + 2 = 9) U =
   9 x 432000 (human) years = 3888000 (human) years =
   3.888 million (human) years

Kali-Yuga by the (human) calendar year 2020 = 5114 (human)
   years =0.005114 million years

Adding   up,   we   get   the   Total   =   (1,851.4285716
   +116.64+3.888+0.005114) million (human) years
   = 1971.9616856 million (human) years
   = 1.9719616856 billion (human) years

This is the cosmic Time-epoch in the 51st Year of Brahmaa's Life.
Hence, the approximate *net* life of Brahmaa in the present
   Maha Kalpa is
   = (155.52 trillion + 1.972 billion) (human) years
   = 155.521972 trillion years.

The Big Bang theory estimates the age of the solar system
to be about 4.3 billion years and the age of the universe to be in
the range of 13.7 billion years. As mentioned, the duration of a
Kalpa (4.32 billion years) coincides with the estimated age of the
earth and the solar system according to the modern cosmology;
while the estimated age of the universe is little over 3 times the
above duration, according to the modern cosmology. The Hindu
scriptural cosmology spans an enormously vast duration of
time due to the periodic origin, evolution, and dissolution of the
universe cyclically and indefinitely.

To recapitulate, in terms of human time-scale:
   Janma Kalpa (one day of Brahmaa or one cycle of creation) =
      4.32 billion years
   Pralaya Kalpa (one night of Brahmaa or one cycle of
      dissolution) = 4.32 billion years
   Maha Kalpa (100 years-age of Brahmaa) =311.04 trillion
      years

Elapsed time in the present Mahaa Kalpa (age of Brahmaa) = 155.52197 trillion years

Creation & Dissolution cycle (Janma Kalpa + Pralaya Kalpa) = 8.64 billion years

Number of creation & dissolution cycles within one life-span of Brahmaa = (311.04 trillion/8.54 billion) = 36421.545 (=36,000)

Manavantara: 14 in one *Janma Kalpa* = (4.32/14) billion years = 308.5714 million year

One Manavantara = 72 Chatuyugees

Number of Chatuyugees in one cycle of creation (or dissolution) = (8.64 billion/4.32 million) = 2000

Chaturyugee (Maha Yuga): Total period of 4 Yugas = 4.32 million years

One Kalpa (Janma or Pralaya) = 1,000 *Chaturyugee (Mahaayuga)*

According to Hindu view as outlined in Shreemad Bhaagavatam the physical universe is created and dissolved periodically and then is reborn. The above cosmic cycle is repeated indefinitely without beginning and without end of Time.

NOTES:

- In the context of modern science, presently, the universe is expanding. All galaxies are receding with respect to one another. One view supported by the presently available data is that the universe may expand forever. The other view is that this universe is restored to its original (beginning) state of very dense mass as postulated in Big Bang theory. However, according to Hindu view the universe is created from a point called *Bindu Vishphot* (the Point of Explosion) that essentially is in agreement with the modern Big Bang theory of cosmology. According to the Hindu scriptures, the universe will revert back to the dense mass referred to as the Big Crunch in the terminology of modern Cosmology. This cyclic pattern is repeated indefinitely, without beginning and without end.

- *The Hindu view of the formation of cosmos is described in Puraanas, which are more than 5000 year old, while the theory of Big Bang has been proposed only recently. The above astounding agreement between the estimates of the age of the universe speaks volumes about the depth of knowledge, on part of the ancient Hindu scholars, about the cosmic time scales and phenomena. A few other facets in this connection are briefly mentioned below.*

- In the *Surya-Siddhaanta*, the Hindu astronomer, Bhaaskaraachaarya, calculates the time taken for the earth to orbit the sun to 9 decimal places and is given equal to 365.258756484 days. It can be compared to the value of 365.2564 days given by Astronomer W.M. Smart and currently accepted measurement of 365.2596 days. Between Bhaskaracharya's ancient measurement 1500 years ago and the modern measurement, the difference is only 0.00084 days, i.e. only 0.0002%.

- The longest unit of time in the history of humanity is the Maha-*Kalpa* in Hindu cosmological Chronology. In astronomy a cosmic year, the period of rotation of the sun around the center of the Milky Way Galaxy, is 0.225 billion years, while the *Kalpa (Janma + Pralaya)* is equivalent to 8.64 billion years. From the largest unit discussed above to the smallest unit of time, *Krati*, which is a measure of 34,000th of a second, Bhaaskaraachaarya gives details about the different measures of time.

APPENDIX B

A Few Great Ancient Indian Scientists

A great number of Indian scientists and philosophers made basic and important contributions in the fields of physics, chemistry, mathematics, astronomy, medicine, surgery, logic, and so on. Many of these contributions preceded the 'discovery' of the pertinent principles and theories by western scientists by centuries, even millennia; although ignorance about the Indian scientific achievements in the west led to the belief that these principles and theories were discovered by the western scientists.

Prominent Indian scientists include **Aryabhatt** (476 BCE), **BHASKARACHARYA II** (1114-1183 CE), **ACHARYA KANAD** (600 BCE), **NAGARJUNA** (100 CE), **ACHARYA CHARAK** (600 BCE), **ACHARYA SUSHRUT** (600 BCE**), VARAHMIHIR (?), ACHARYA PATANJALI** (200 BCE**), ACHARYA BHARADWAJ** (800 BCE), and **ACHARYA KAPIL** (3000 BCE), among others. Brief accounts of their specific contributions are provided below. A few 'Artist's renditions are presented at the end.

ARYABHATT
(476 CE)
MASTER ASTRONOMER AND MATHEMATICIAN

Born in 476 CE in Kusumpur (Bihar), Aryabhatt's intellectual brilliance remapped the boundaries of mathematics and astronomy. In 499 CE, at the age of 23, he wrote a text on astronomy and an unparallel treatise on mathematics called "Aryabhatiyam." He formulated the process of calculating the motion of planets and the time of eclipses. Aryabhatt was the first to proclaim that the earth is round, it rotates on its axis, orbits the sun and is suspended in space—1000 years before Copernicus published his heliocentric theory. He is also acknowledged for calculating π (Pi) to four decimal places: 3.1416 and the sine table in trigonometry. Centuries later, in 825 CE, the Arab mathematician, Mohammed Ibna Musa credited the value of Pi to the Indians, "This value has been given by the Hindus." And above all, his most spectacular contribution was the concept of zero without which modern computer technology would have been non-existent. Aryabhatt was a colossus in the field of mathematics.

BHASKARACHARYA II
(1114-1183 CE)
GENIUS IN ALGEBRA

Born in the obscure village of Vijjadit (Jalgaon) in Maharastra, Bhaskaracharya's work in Algebra, Arithmetic and Geometry catapulted him to fame and immortality. His renowned mathematical works called "Lilavati" and "Bijaganita" are considered to be unparalled and a memorial to his profound intelligence. Its translation in several languages of the world bear testimony to its eminence. In his treatise "Siddhant Shiromani" he writes on planetary positions, eclipses, cosmography, mathematical techniques and astronomical equipment. In the "Surya Siddhant" he makes a note on the force of gravity:

"Objects fall on earth due to a force of attraction by the earth. Therefore, the earth, planets, constellations, moon, and sun are held in orbit due to this attraction." Bhaskaracharya was the first to discover gravity, 500 years before Sir Isaac Newton. He was the champion among mathematicians of ancient and medieval India. His works fired the imagination of Persian and European scholars, who through research on his works earned fame and popularity.

ACHARYA KANAD
(600 BCE)
FOUNDER OF ATOMIC THEORY

As the founder of "Vaisheshik Darshan"—one of six principal philosophies of India—Acharya Kanad was a genius in philosophy. He is believed to have been born in Prabhas Kshetra near Dwarika in Gujarat. He was the pioneer expounder of realism, law of causation and the atomic theory. He has classified all the objects of creation into nine 'elements', namely: earth, water, light, wind, ether, time, space, mind and soul. He says, "Every object of creation is made of atoms which in turn connect with each other to form molecules." His statement ushered in the Atomic Theory for the first time ever in the world, nearly 2500 years before John Dalton. Kanad has also described the dimension and motion of atoms and their chemical reactions with each other. The eminent historian, T.N. Colebrook, has said, "Compared to the scientists of Europe, Kanad and other Indian scientists were the global masters of this field."

NAGARJUNA
(100 CE)
WIZARD OF CHEMICAL SCIENCE

He was an extraordinary wizard of science born in the nondescript village of Baluka in Madhya Pradesh. His dedicated research for twelve years produced maiden discoveries and inventions in the faculties of chemistry and metallurgy. Textual masterpieces like "Ras Ratnakar," "Rashrudaya" and "Rasendramangal" are his renowned contributions to the science of chemistry. Where the medieval alchemists of England failed, Nagarjuna had discovered the alchemy of transmuting base metals into gold. As the author of medical books like "Arogyamanjari" and "Yogasar," he also made significant contributions to the field of curative medicine. Because of his profound scholarliness and versatile knowledge, he was appointed as Chancellor of the famous University of Nalanda. Nagarjuna's milestone discoveries impress and astonish the scientists of today.

ACHARYA CHARAK
(600 BCE)
FATHER OF MEDICINE

Acharya Charak has been crowned as the Father of Medicine. His renowned work, the "Charak Samhita", is considered as an encyclopedia of Ayurveda. His principles, diagoneses, and cures retain their potency and truth even after a couple of millennia. When the science of anatomy was confused with different theories in Europe, Acharya Charak revealed through his innate genius and enquiries the facts on human anatomy, embryology, pharmacology, blood circulation, and diseases like diabetes, tuberculosis, heart disease, etc. In the "Charak Samhita" he has described the medicinal qualities and functions of 100,000 herbal plants. He has emphasized the influence of diet and activity on mind and body. He has proved the correlation of spirituality and physical health, and contributed greatly to diagnostic and curative sciences. He has also prescribed the ethical charter for medical practitioners two centuries prior to the Hippocratic oath. Through his genius and intuition, Acharya Charak made landmark contributions to Ayurveda. He forever remains etched in the annals of history as one of the greatest and noblest of Rishi (seer)-scientists.

ACHARYA SUSHRUT
(600 BCE)
FATHER OF PLASTIC SURGERY

A genius who has been glowingly recognized in the annals of medical science. Born to sage Vishwamitra, Acharya Sudhrut details the first ever surgery procedures in "Sushrut Samhita," a unique encyclopedia of surgery. He is venerated as the father of plastic surgery and the science of anesthesia. When surgery was in its infancy in Europe, Sushrut was performing Rhinoplasty (restoration of a damaged nose) and other challenging operations. In the "Sushrut Samhita," he prescribes treatment for twelve types of fractures and six types of dislocations. His details on human embryology are simply amazing. Sushrut used 125 types of surgical instruments including scalpels, lancets, needles, Cathers and rectal speculums; mostly designed from the jaws of animals and birds. He has also described a number of stitching methods; the use of horse's hair as thread and fibers of bark. In the "Sushrut Samhita," he details 300 types of operations. The ancient Indians were the pioneers in amputation, caesarian and cranial surgeries. Acharya Sushrut was a giant in the arena of medical science.

VARAHMIHIR
EMINENT ASTROLOGER AND ASTRONOMER

Renowned astrologer and astronomer who was honored with a special decoration and status as one of the nine gems in the court of King Vikramaditya in Avanti (Ujjain). Varahamihir's book "Panchsiddhant" holds a prominent place in the realm of astronomy. He notes that the moon and planets are lustrous not because of their own light but due to sunlight. In the "Bruhad Samhita" and "Bruhad Jatak," he has revealed his discoveries in the domains of geography, constellation, science, botany, and animal science. In his treatise on botanical science, Varahmihir presents cures for various diseases afflicting plants and trees. The rishi-scientist survives through his unique contributions to the science of astrology and astronomy.

ACHARYA PATANJALI
(200 BCE)
FATHER OF YOGA

The Science of Yoga is one of several unique contributions of India to the world. It seeks to discover and realize the ultimate Reality through yogic practices. Acharya Patanjali, the founder, hailed from the district of Gonda (Ganara) in Uttar Pradesh. He prescribed the control of prana (life-force) as the means to control the body, mind, and soul. This subsequently rewards one with good health and inner happiness. Acharya Patanjali's 84 yogic postures effectively enhance the efficiency of the respiratory, circulatory, nervous, digestive, and endocrine systems and of many other organs of the body. Yoga has eight limbs where Acharya Patanjali shows the attainment of the ultimate bliss of God in samadhi through the disciplines of: yam, niyam, asan, pranayam, pratyahar, dhyan and dharna. The Science of Yoga has gained popularity because of its scientific approach and benefits. Yoga also holds the honored place as one of six philosophies in the Indian philosophical system. Acharya Patanjali will forever be remembered and revered as a pioneer in the science of self-discipline, happiness and self-realization.

ACHARYA BHARADWAJ
(800 BCE)
PIONEER OF AVIATION TECHNOLOGY

Acharya Bharadwaj had a hermitage in the holy city of Prayag and was an ardent apostle of Ayurveda and mechanical sciences. He authored the "Yantra Sarvasva" which includes astonishing and outstanding discoveries in aviation science, space science, and flying machines. He has described three categories of flying machines: 1.) One that flies on earth from one place to another; 2.) One that travels from one planet to another; and 3.) One that travels from one universe to another. His designs and descriptions have impressed and amazed aviation engineers of today. His brilliance in aviation technology is further reflected through techniques described by him:

1.) Profound Secret: The technique to make a flying machine invisible through the application of sunlight and the wind force.
2.) Living Secret: The technique to make an invisible space machine visible through the application of electrical force.
3.) Secret of Eavesdropping: The technique to listen to a conversation in another plane.
4.) Visual Secrets: The technique to see what's happening inside another plane.

Through his innovative and brilliant discoveries, Acharya Bharadwaj has been recognized as the pioneer of aviation technology.

ACHARYA KAPIL
(3000 BCE)
FATHER OF COSMOLOGY

Celebrated as the founder of Sankhya philosophy, Acharya Kapil is believed to have been born in 3000 BCE to the illustrious sage Kardam and Devhuti. He gifted the world with the Sankhya School of Thought. His pioneering work threw light on the nature and principles of the ultimate Soul (Purusha), primal matter (Prakruti) and creation. His concept of transformation of energy and profound commentaries on atma, non-atma and the subtle elements of the cosmos places him in an elite class of master achievers—incomparable to the discoveries of other cosmologists. On his assertion that Prakruti, with the inspiration of Purusha, is the mother of cosmic creation and all energies, he contributed a new chapter in the science of cosmology. Because of his extrasensory observations and revelations on the secrets of creation, he is recognized and saluted as the Father of Cosmology.

MATHEMATICS

Rishi - Scientists of India

ALGEBRA

ATOMIC THEORY

Rishi - Scientists of India

COSMOLOGY

Rishi - Scientists of India

CHEMISTRY

"Philosophers have said that if the same circumstances don't always produce the same results, predictions are impossible and science will collapse.....Physics, a science of great exactitude, has been reduced to calculating only the probability of an event, and not predicting exactly what will happen. ...That's a retreat, but that's the way it is. Nature permits us to calculate only probabilities. Yet science has not collapsed."

Richard Phillips Feynman (1918-1988)

www.ingramcontent.com/pod-product-compliance
Lightning Source LLC
Chambersburg PA
CBHW032003170526
45157CB00002B/523